ELECTRICITY

David Bryant has wide experience in teaching and lecturing in physics, and is now Principal Adviser for Secondary Education, County of Dorset. He is author of *Teach Yourself Physics* and co-author of *A New Physics*.

TEACH YOURSELF BOOKS

ELECTRICITY

David Bryant

TEACH YOURSELF BOOKS

Hodder and Stoughton

First published 1987

British Library Cataloguing in Publication Data
Bryant, D. (David), *1936–*
Electricity. – (Teach Yourself books)
1. Electricity
I. Title
537   QC522

ISBN 0 340 40713 1

Printed and bound in Great Britain
for Hodder and Stoughton Educational,
a division of Hodder and Stoughton Ltd,
Mill Road, Dunton Green, Sevenoaks, Kent
by Richard Clay Ltd,
Bungay, Suffolk. Photoset by
Rowland Phototypesetting Ltd,
Bury St Edmunds, Suffolk

Contents

Introduction

Electricity has often been thought of as one of the more difficult topics in science, from elementary work in schools to advanced studies in universities. The aim of this book is to try to make the subject understandable and approachable to readers who want to feel comfortable when dealing with electrical matters. The areas covered range from basic ideas about current and voltage to aspects of electronics and the distribution of electrical energy. Explanations are pitched at the level appropriate to first examinations, occasionally above that, sometimes below, but the coverage has not been tied to the examination requirements of any one particular syllabus.

Some mathematics is unavoidable, but no great competence is needed to cope with the majority of the calculations and formulae in this book. Readers who wish to pursue topics in more detail will find many standard textbooks which offer deeper insights into the subject. The grounding provided here will give an essential foundation for more advanced study as well as a general framework for those who go no further.

The book uses metric (SI) units almost exclusively. The lists overleaf give the quantities, units and abbreviations and the multiples and submultiples most commonly met in practical situations. In the text, important equations, laws and relationships are all clearly highlighted. Conventional signs and symbols follow almost completely those listed in British Standard 3939.

Quantity	Unit	Abbreviation
length	metre	m
area	square metre	m^2
volume	cubic metre	m^3
mass	kilogram	kg
time	second, hour	s, h
velocity	metre per second	ms^{-1}
force	newton	N
current	ampere	A
potential difference	volt	V
resistance	ohm	Ω
energy or work	joule	J
	kilowatt hour	kWh
power	watt	W
charge	coulomb	C
capacitance	farad	F
inductance	henry	H
frequency	hertz	Hz

Prefix	Abbreviation	Value	Example
mega	M	million	1 MΩ = 1 000 000 ohms
kilo	k	thousand	1 kW = 1000 watts
centi	c	hundredth	$1 \text{ cm} = \dfrac{1}{100}$ metre
milli	m	thousandth	$1 \text{ mA} = \dfrac{1}{1000}$ amp
micro	μ	millionth	$1 \text{ μC} = \dfrac{1}{1\,000\,000}$ coulomb
pico	p	million millionth	$1 \text{ pF} = \dfrac{1}{1\,000\,000\,000\,000}$ coulomb

Safety and Practical Work

Most of the risks associated with the handling of electrical appliances or installations can be avoided by commonsense precautions. It is the mains supply which brings the greatest danger because its high voltage and the current it can deliver are potentially lethal in the right circumstances. Low voltages, however, such as those used in battery driven devices like transistor radios and toys, present no risks at all. The worst that could happen with these would be a short circuit across the battery which would quickly exhaust it, or a wrong connection which might damage a component but not a person.

The two main dangers from the 240 V mains supply are *shocks* and *overloading*. Shocks are felt when a current passes through the body from a source of about 60 V or more, capable of delivering a current of anything over ½ A. People differ greatly in their sensitivity to electric shocks and in their tolerance of them, so it is sensible to treat the mains sockets, switches and appliances with very great care. Overloading can occur by too liberal use of adaptors, which might result in more than the safe current passing through cables in the house. The cables then overheat and perhaps cause a fire.

The following precautions should be taken with mains electricity:

DO   — make sure the supply is switched off before dismantling or examining equipment or appliances;
      — wear shoes with insulating soles;
      — call in qualified help if in any doubt;
      — check connections for tightness and for stray strands of wire;
      — use the colour code for wiring plugs and sockets;
      — use the correct size of fuse for each appliance and circuit.

DON'T — remove protective covers from equipment still connected to the mains;
— drill into walls directly above or below light switches or sockets;
— use adaptors without checking the total current demand;
— use lighting circuits for heating or power;
— use twin cable for appliances which need earthing.

(The principles behind many of these points are dealt with in various parts of the book, particularly Chapter 13 for safety in the home.)

Practical work is often the best way to develop real insight into a topic like electricity. For those who wish to try some of the experiments and arrangements described in this book the following list of items, which are reasonably easy to obtain, is suggested. Some other items lend themselves to improvisation without much difficulty.

1.5 V dry cells	simple multimeter
4.5 V dry battery	magnets
cell holder	iron filings
small lamp holders	plotting compass
1.25 V and 4.5 V lamps	soft iron rods
crocodile clips	selection of: diodes
connecting wire (single core, plastic coated)	resistors transistors
wire strippers	thermistor
low voltage switches	selection of fuses and fuse wire
variable resistor 0–10 Ω	

For more serious work, especially in electronics, there are several forms of board available on which more permanent circuits can be mounted. Specialist magazines carry advertisements for these, together with many other items of electrical or electronic equipment.

1

Electric Current

1.1 Current means flow

Many branches of science have become full of special words, a sort
of jargon which often gets in the way of understanding exactly what
is happening. There are some in electricity which will be met later
but for the basic idea of an electric *current* we use the word in a way
similar to everyday language. People talk about a current of air or a
tidal current or a river current or even a current of opinion when
they mean that the air or the tide or the river or public opinion is
moving or *flowing* along in a particular direction. An electric
current implies just the same thing – a flow of some kind – and it is
reasonable to expect that the flow of electricity is similar in some
ways to the flow of air or water, as indeed it is.

1.2 A flow of what? Electric charge

It is easy to understand air or water currents since they can be felt
directly by standing in the wind or putting a hand into a stream, or
their effects can be observed through the movements of clouds or
the drifting of a boat. There really is air or water moving along, with
clear effects for all to see. A current of opinion is rather more
difficult to understand because there is nothing visible to watch
moving or flowing. Yet the spread of an idea or a fashion across a
whole country is real enough and its effects can be noticed very
clearly.

In the case of an electric current it is not possible to see the actual
thing which is flowing – only the effects of the flow. Even a lightning

flash is the consequence of a very sudden flow of electricity, not the flow itself. An electric current is a flow of electric *charge*, carried by tiny particles called *electrons* or *ions*, which can be either *negative* or *positive*. So an electric current can best be described as a flow of electrons or a flow of ions. As a rule, it is electrons that move when electricity flows through solid materials but ions when it flows through liquids or gases.

1.3   How much flow? Coulomb per second

The charges carried by electrons and ions are very small, as the particles are themselves. An electron has a mass of around 10^{-30} kg whilst an ion is much larger. An ion may have the same charge as an electron, or exactly two or three times as much.

Charge is measured in *coulombs*. An electron carries only a tiny fraction (1.6×10^{-19}) of a coulomb, so in fact about 6×10^{18} electrons (six million million million) together have a charge of one coulomb. When charges move from place to place, i.e. when electrons or ions are caused to move, the *flow* of electricity is measured in coulombs per second rather than electrons or ions per second.

This measure of electric current as coulombs per second corresponds exactly with ways of describing other currents: for example, a river flows at a certain number of cubic metres per second, or traffic moves along a motorway at a rate of so many thousands of vehicles per hour, or water is pumped along a pipeline at a rate of so many millions of cubic metres per day. Thus, an electric current, like other flows, is measured by the rate at which the basic quantity moves – volume of liquid, or vehicles, or in this case electrons with their charge. The basic unit of a coulomb per second implies about 10^{18} electrons per second.

1.4   What is an ampere?

To complete the picture, a special name is given to the unit in which electric current is measured: the *ampere*, which is exactly the same as a coulomb per second, just a shorter name for the same thing.

---

1 ampere $\equiv$ 1 coulomb per second
1 ampere means a flow of $\sim 10^{18}$ electrons per second

The unit is often further abbreviated to amp or just simply A. A current of 5 A means five amperes (amp) or five coulombs per second. The names come from those of André Marie Ampère (died 1836) who worked on the relationship between electricity and magnetism, and Charles Augustin Coulomb (died 1806) who investigated forces between charges. (See also section 9.2.)

1.5 Sizes of electric currents

To help to get a feeling for the size of an ampere it is useful to know the sizes of currents in everyday use. Since the ampere is quite a large unit for many purposes we also use the milliamp (mA, a thousandth of an amp) instead. An even smaller unit, the micro amp (μA, a millionth of an amp), is occasionally needed.

Here are some common devices and the currents typically used to operate them:

Shaver	30 mA	Video recorder	1.0 A
Transistor radio	0.2 A	Refrigerator	1.5 A
Torch	0.3 A	Drill	2.0 A
100-watt lamp	0.4 A	Home computer	3.0 A
Model train	0.5 A	Iron	3.0 A
Music centre	0.6 A	Toaster	6.3 A
Television	0.8 A	Immersion heater	12 A

1.6 Speed of electrons' movement

Although a current of 1 A means some 10^{18} electrons passing in a second, and although the effects of the current can be seen almost instantaneously, the rate at which electrons actually move along a wire is surprisingly slow. The speed depends on the size of the current, as well as the diameter and material of the wire: if we have 1 A flowing in a copper wire of 1 mm diameter, it turns out that the electrons drift along at only about $\frac{1}{10}$ mm per second. The larger the current the greater is the speed, but even in high current devices it is still a surprisingly slow movement.

1.7   Direction of flow of electrons and electric current

In all circuit diagrams, such as those which will be developed in later chapters and which are found with certain electrical appliances (such as transistor radios) and in motor car handbooks, it is always assumed that the current passes from positive to negative, + to −. In fact it is now believed that the electrons actually move in the opposite direction, from − to +. When scientists were developing the early ideas of electricity the electron had not been discovered and the names positive and negative had already been given to the two kinds of charge. Later the electron was found to have a negative charge.

When current passes, then, it is the extra electrons at the negative terminal which are moved through the connecting wires to the positive terminal of the supply, i.e. from − to +. Unfortunately, before this was fully understood the habit had developed of describing electric currents as passing from + to −, the so-called *conventional* direction. In practice, the difference does not affect many aspects of studying or understanding electricity, but it is as well to be aware of the historical accident which has resulted in a 'wrong' description. (It is only in certain applications of electronic devices where the actual electron-flow has to be considered rather than the conventional flow.)

1.8   Ammeters

Any device which measures the size of an electric current is called an *ammeter*. There are several ways of constructing such a device, which will be considered later (Chapters 6 and 9), but the commonest kind indicates the current, i.e. the rate of flow of electric charge, by a steady reading of a needle on a simple scale marked in amps or milliamps. The ammeter is placed so that the current to be measured actually passes through the ammeter (Figure 1.1).

A well designed ammeter will show the current clearly and accurately, and in many circuits the instrument would be made a permanent feature of the installation, always able to give a visual indication of the current − for example, in a battery charger. Sometimes, though, it is necessary to introduce an ammeter into a circuit temporarily, to see what current is passing − when testing a

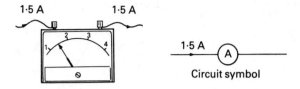

Fig. 1.1 An ammeter measures the current flowing through it

transistor radio, for example. In this case, it is important for the ammeter to show the current which was passing before it was connected into the circuit – the ammeter itself should not cause the size of the current to change because of its own internal design. We talk about an ammeter measuring the current *in* or *through* a wire or component.

1.9 Direct and alternating current

The idea of electric current has been introduced as a flow of charge carried by electrons travelling in a particular direction. A time-graph of such a current looks like Figure 1.2 – always in the same direction and usually steady in size.

We call this a *direct current*, often abbreviated to d.c. All battery-operated devices work on d.c. and it will be used in much of this book to develop the main properties of electricity, partly because it is easy to visualise what is happening.

There is another important way in which electricity can flow, called *alternating current* or a.c. Rather than passing always in one direction, an alternating current passes first one way then the other,

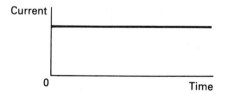

Fig. 1.2 Time-graph of a direct current

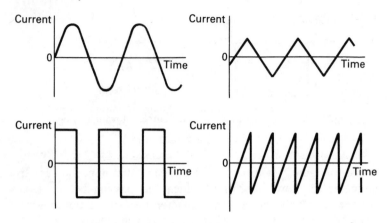

Fig. 1.3 Time-graphs of various alternating currents

forwards and backwards, continually changing its direction. Time-graphs of different examples of a.c. are shown in Figure 1.3.

The shape of the current-time graph depends on how the a.c. is generated. The smoothly changing a.c. in the first example is the commonest one, used for the mains electricity supply to houses, offices and factories. Alternators in motor cars and cycle dynamos also produce alternating currents. We shall meet some properties and uses of a.c. which differ from d.c., but many of the elements of electricity apply equally to both.

Using the water-flow analogy introduced at the start of the chapter, the movement of water through a central heating system is a good example of a direct current, whilst the large scale ebb and flow of the tides constitutes an alternating current which fluctuates almost twice each day.

1.10   Summary

An electric current is a flow of charge carried by electrons or ions.

Charge is measured in coulombs and current in amperes.

An ampere is a coulomb per second.

Current is said to pass from + to − around a circuit.

The size of a current is indicated by an ammeter, which itself becomes part of the circuit carrying the current.

Direct current is always in the same direction.

Alternating current passes forwards and backwards in turn.

2

Simple Circuits

2.1 Current passing round a loop

Electric circuits, like central heating water circuits, in their simplest form contain three main parts: (*a*) a continuous loop of material through which the current can pass, (*b*) a device which is able to move the charges (or the water) around the loop, and (*c*) a place where the current does useful work for us. In an electric circuit, charges (electrons) flow through a continuous loop of wires, moved on their way by a cell (e.g. torch 'battery'), and cause a lamp to light or a motor to operate. In a central heating circuit hot water flows through a loop of pipes, moved along by a pump, and heats up a radiator to warm the air around it (Figure 2.1).

It is important to realise that there must be a *closed* loop of wire or water pipes to maintain the flow all the way round. It is only when both wires are connected between lamp and cell that the lamp lights; just one wire will not do, because it is the flow of the electrons

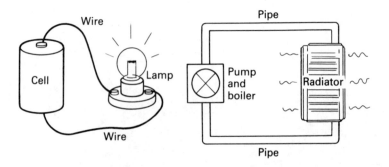

Fig. 2.1 Electric and water circuits

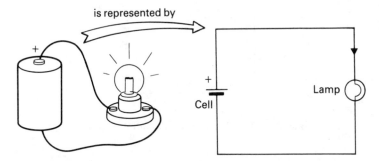

Fig. 2.2   Current direction is indicated from + to −

into, through and out of the lamp which causes the filament to become hot. Similarly with the radiator – a continuous flow of warm water is needed to keep the radiator hotter than the surrounding air. (There is always a difficulty if an analogy is taken too far – in the water circuit there must also be a heater or boiler in addition to the components shown to make the water hot.) If for any reason the closed loop were broken, by disconnecting a wire or switching the pump off, for instance, the flow would stop, making the lamp go out and the radiator cool down.

2.2   Direction of flow

As explained in the previous chapter (section 1.7), we show electric currents passing from the + terminal of the cell through the connecting wires and appliances to the − terminal, even though we know that electrons (the things that actually move around the circuit) flow the other way. In circuit diagrams, the usual practice is to draw them showing the current from top to bottom, through the appliances in which energy is transferred – lamps, motors, coils, heaters, etc. (Figure 2.2).

2.3   Size of current around the circuit

An electric current is a flow of charge and the size of the current indicates the rate of flow. A simple experiment using two ammeters (or the same ammeter in two different places) will show the sizes of the currents going into and leaving the lamp (Figure 2.3).

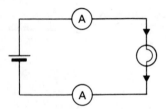

Fig. 2.3   Measuring current at different points

You may be surprised to learn that the ammeters indicate the same current entering and leaving the lamp. It is a common error to think that the current gets 'used up' in the lamp, with more entering it than leaving it, but if the idea of *rate of flow* is understood it is not possible for any current to disappear unless it somehow 'leaks' out of the lamp. (Think of the water flow through a radiator.)

What does happen inside the lamp is that electrical energy generated by the cell is transferred to thermal and light energy, just as in a radiator energy from the boiler is transferred to the air in the room.

The current is the same everywhere around a simple circuit (Figure 2.4), no matter where it is measured nor how many components there are – even inside the cell, if only an ammeter could be placed there to measure it. Again, the water circuit, in which the flow has to go through the pump as well as everything else, offers a good parallel.

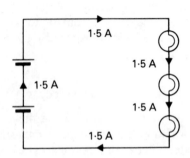

Fig. 2.4   Current is the same everywhere in a closed loop

2.4 Currents at junctions in circuits – Kirchhoff's Law

Very few circuits in electrical appliances are in fact simple loops, so we need to consider what happens to the current where junctions occur. Again, the basic idea of rate of flow gives the answer without difficulty. If, say, six coulombs per second flow into a junction the same number must flow out of it per second, otherwise there would be a surplus of electrons building up at the junction or a mysterious disappearance of them (Figure 2.5). (A similar result would be true of water flow, of course, assuming no leaks!)

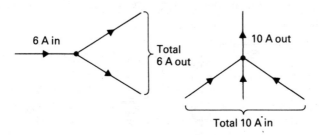

Fig. 2.5 Currents at a junction

Generalising further than the two simple cases illustrated in Figure 2.5, we can say that the same is true of more complex junctions also (Figure 2.6).

The relationship between electric currents at junctions is summarised by Kirchhoff's Law, which states that:

> The total current entering any junction in an electrical network equals the total leaving the junction.

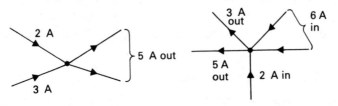

Fig. 2.6 Currents into a junction equal those out

In the two examples in Figure 2.6 many different possibilities exist, all of which satisfy Kirchhoff's Law: the two currents in the first might be 4 A and 1 A, or 2.5 A and 2.5 A, or 0 A and 5 A or even 6 A out and 1 A in. The reader can think of many similar cases for the second example.

2.5   Connecting in series

The simple loop circuit of Figure 2.2 has the same current through all the parts of the circuit. It is a simple arrangement in which there are no junctions or branches, just one part connected to the other in a chain. A circuit in which this happens is called a *series* circuit (Figure 2.7).

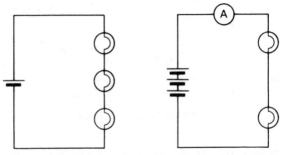

Fig. 2.7   Examples of series circuits

The word *series* is applied to the whole circuit as well as to the way the lamps or the cells are connected to each other. Thus the lamps are connected 'in series' in the first circuit, and the cells in the second. Further, in each circuit the cell(s) and the lamp(s) are also in series. Any chain-like connection of components through which the same current passes is called a *series* connection, or the components are said to be connected *in series*.

One feature of a series circuit, such as a set of Christmas tree lights, is that since the current passes through each lamp in turn the failure of a single lamp causes them all to go out, because the circuit is broken.

2.6   Connecting in parallel

When elements in a circuit are arranged in such a way that the current divides into two or more alternative paths, which subsequently reunite, they are said to be connected *in parallel* with one another. One feature of this arrangement is that the total current is shared between the parallel paths at a junction and restored to its original value at a second junction (Figure 2.8).

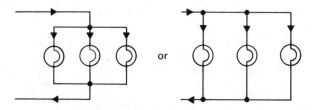

Fig. 2.8   Connections in parallel

Twin rear lamps on a car are an example of a parallel connection, as are the sockets in a house wiring system and lamps in a chandelier. In this way of connecting lamps together, the failure of one lamp does not cause them all to go out.

2.7   Practical work with simple circuits

The properties of series and parallel circuits can be explored with a set of cells, lamps and wires by arranging circuits such as those shown in Figure 2.9. Doing this would also give the reader a feeling for what factors lead to larger and smaller currents, using the brightness of the lamps as indicators of current size.

2.8   Mixed series and parallel connections

The majority of circuits in practical applications are a mixture of series and parallel connections, and might even contain more than one source of electricity. It is more usual in complicated arrangements to use the words 'series' and 'parallel' for *groups* of components rather than for the circuit as a whole. For example, in Figure 2.10, the cell and the main switch are in series whilst the three pairs of lamps and switches are in parallel with one another.

The circuit diagram of a simple transistor amplifier further illustrates how series and parallel connections can be combined (Figure 2.11).

Series

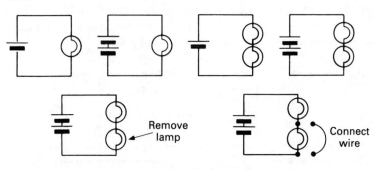

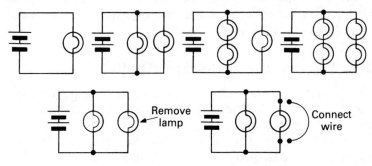

Fig. 2.9   Investigating series and parallel circuits

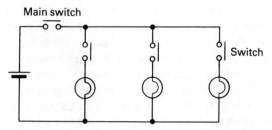

Fig. 2.10   Mixed series and parallel connections

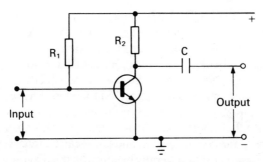

Fig. 2.11 Series and parallel connections in an amplifier circuit

2.9 Circuit diagrams and symbols

The list which follows includes most of the symbols for circuit diagrams used in this book:

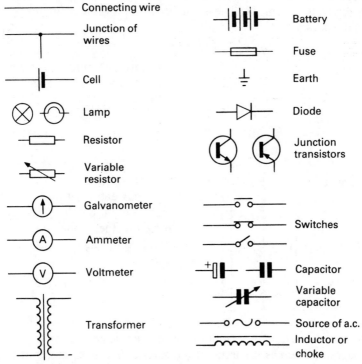

Symbols may be shown in almost any position or orientation.

The simplest circuit diagram has been met already – a single loop with no branches or junctions – and also a few of the commonest symbols – cell, lamp, connecting wire. Many more will appear throughout the book as a convenient shorthand for a variety of practical circuits. Seldom will the diagrams reflect the exact positions of the components in practice, but they show the relationships between components in a clear way. How the circuits work can be understood much more easily from such diagrams than from the appearance of the actual devices, and faults can be traced systematically too. Circuit design generally begins with the diagrams and then develops towards a practical layout.

2.10   Currents in series and parallel arrangements

From what has been said so far it is clear that two straightforward results follow, which will be useful in considering other circuits later in the book:

Series

The current through each component in a series arrangement is the same.

Parallel

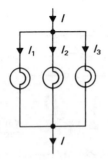

The currents through each branch of a parallel arrangement together equal the total current entering the arrangement:

$$I = I_1 + I_2 + I_3$$

Notice the use of the symbol I for an electric current on a circuit diagram, whereas the letter A indicates either an ammeter or the unit of current, the ampere.

2.11 Summary

Electric currents pass round loops of conducting materials called circuits.

The size of the current is the same at all points around a simple loop.

At a junction in a circuit the total current entering it is the same as that leaving it.

In a series connection the same current flows through each component.

In a parallel connection different currents may pass through each component.

Standard symbols are used to represent components in circuit diagrams.

3

Voltage

3.1 Electrical 'force'

The flow of an electric charge has been likened to the movement of water through pipes: for both, it is natural to talk about a current as the rate of flow of the thing that moves. Keeping to the analogy, we can see in the case of water that it will flow around the system of pipes only if there is a pump to push it along. What the pump does is to create a force which keeps the water moving through the pipes in a continuous flow. The current of water, i.e. its rate of flow, depends on there being a force to cause the movement – with the pump switched off and the force removed, the current is zero.

Similarly in the electrical case, a flow of electrons (i.e. a current) depends on the generation of an electrical *force* to cause the movement of electrons along the wires and through the lamps or other circuit elements. It is the cell which provides this force in the simple circuit by virtue of the chemical action taking place within it. The term used is *electromotive force*, which describes exactly the effect it has – it forces electric charges to move.

3.2 Electrical 'level'

Another way of looking at the same ideas, in the water analogy, is to say that a pump produces a difference of pressure between its inlet and outlet which causes water to move from the high pressure outlet to the low pressure inlet when a pipe is connected between them. In other words, the pressure difference causes the flow of water. (Pressure differences in the atmosphere similarly cause air to flow from place to place – the winds.) In the same way, thermal energy will flow along a metal bar if there is a temperature difference between its ends. These similar arrangements all have in common a difference of 'level' of some property which is able to set in motion a flow of some kind.

In electricity the same kind of language is used: the terminals of a cell are thought of as being at different electrical 'levels', one higher than the other. Charge flows when the different levels are connected together with an appropriate material. Both the 'level' and 'force' descriptions are useful because they link the flow of electricity with other types of flow in a common language, illustrating the close similarities between them.

3.3 Electromotive force, potential difference and voltage

Electromotive force (section 5.4) is developed by a cell or any source of electrical energy such as a generator or dynamo, and is measured in *volts* (named after the Italian scientist Alessandro Volta, d. 1827). The difference in electrical 'level', called *potential difference*, between two points in a circuit is also measured in volts. The two ideas of 'force' and 'level' are merely alternative ways of describing what it is that causes charge to flow, so it is sensible to measure them in the same units.

When an e.m.f. or a p.d. (electromotive force or potential difference) causes electric charge to flow, which in turn can light a lamp or drive a motor, there is clearly an energy transfer from the source of the electricity to the lamp or motor. Chemical energy in the cell, for example, is changed to thermal and light energy in the lamp. As in other branches of physics, when energy is transferred from one object to another, work has to be done to effect the change. The amount of work equals the amount of energy transferred and both are measured in *joules*. (You would do almost one joule of work in lifting a 100 g mass vertically through a distance of one metre on Earth, and in so doing would transfer one joule of energy from your body to the object you lifted.) The unit of e.m.f. or p.d., the volt, is therefore related to the amount of work needed to move the electrons (or charge) which constitute the current: i.e. the volt must be related to the joule and the coulomb. We will simply note at this stage that a volt is a joule per coulomb and can be regarded as a measure of electrical 'level'. Chapters 5, 8 and 14 deal with the idea in more detail. In the same way, the unit of speed is related to those of distance (metre or kilometre) and time (second or hour) but is not in this case given a special name.

3.4   Volts: + −

If the idea of a difference in electrical 'level' is accepted as so many volts, there is still the matter of whether the difference is upwards or downwards, and of what or where is zero. Contour lines on a large scale map present a similar situation – the difference in height is easy to see, but it takes a close look to decide whether the ground slopes upwards or downwards, and to work out the height above or below sea level. For many purposes in electricity it is sufficient to know what the potential difference is and which of the two points concerned is at the higher 'level', the absolute values for each point not being important.

The positive (+ve) terminal of a cell is said to be at a higher *potential* (level) than the negative (−ve) terminal – in fact the + and − signs are helpful in this case (compare section 1.7). With this convention, electric current passes from high potential to low potential, which fits well with water flowing from high to low pressure and thermal energy from high to low temperature. (It is, of course, the conventional current, not the electron flow!) Conversely, if an electric current is seen to pass between two points in a certain direction, we can easily deduce which point is at the higher potential.

3.5   Zero of potential

Occasionally it is necesary to refer voltages (potentials or 'levels') to a zero level and to know the values of them either above or below that zero. There are two ways in which this arbitrary zero can be defined, a theoretical one and a practical one. The theoretical way is to say that a place remote from any charged objects (infinitely far away!) will be said to be at zero potential, a useful idea for mathematical scientists but not easy to realise in ordinary life.

The practical way is to call the potential of the Earth zero. This is similar to the equally arbitrary notion of mean sea level which is used as a zero reference mark for heights on land and depths in oceans. The large volume of the sea can accommodate water being added or removed without its average level being noticeably affected, making a zero level acceptable as a basis for measurements. Similarly, such a large object as the Earth will not have its electrical level changed noticeably by the addition or removal of even millions of electrons, so it can serve in practice as the zero of potential.

3.6 Earthing

When dealing with electrical safety (Chapter 13) we talk about the case of an appliance such as an electric heater being 'earthed', which means literally connected by a wire to the Earth. One reason for doing this is to have the object and the person who handles it at the same potential, i.e. earth potential or zero, so that no current can pass from one to the other and the person will not get a shock by touching the casing of the appliance.

In electrical circuits, too, it is often useful to earth particular points to bring their potentials to a known value, i.e. zero, which enables potentials at other points to be worked out or specified. The metal chassis on which the circuit is mounted serves as a useful earthing point for this purpose.

3.7 Sizes of electrical voltages

Everyday experiences with electrical equipment of various kinds involve a range of voltages from very small to very large. Multiple and sub-multiple units based on the volt are frequently used to avoid having to write strings of noughts, in the same way as in other branches of physics (see also section 1.5):

$$1 \text{ kilovolt } (1 \text{kV}) = 1000 \text{ volts} \qquad (1000 \text{ V or } 10^3 \text{ V})$$
$$1 \text{ megavolt } (1 \text{ MV}) = 1\,000\,000 \text{ volts} \qquad (1\,000\,000 \text{ V or } 10^6 \text{ V})$$
$$1 \text{ millivolt } (1 \text{ mV}) = \frac{1}{1000} \text{ volt} \qquad \left(\frac{1}{1000} \text{ V or } 10^{-3} \text{ V}\right)$$

Here are some electrical devices and the potential differences which are developed by them or used to operate them:

Torch bulb	1.5 V	Combing dry hair	1 kV
Car headlamp	12 V	TV picture tube	2 kV
Mains supply	240 V	Piezo-electric gas igniter	2 kV
Transistor radio	6 V	Motor car spark plugs	5 kV

3.8   Voltmeters

The measurement of potential differences is most easily done by using a *voltmeter*, which is made in the same way as an ammeter but adapted to measure volts instead of amps. How it actually works will be dealt with in Chapter 6. The important differences between a voltmeter and an ammeter are that the voltmeter is designed to have a high resistance (Chapter 5) and is connected into a circuit in a different way.

Whereas an ammeter is introduced directly into a circuit, such that the current it measures passes through the instrument itself (section 1.8), the voltmeter is an add-on instrument which is connected to the two points whose potential difference is to be measured (Figure 3.1). A voltmeter can thus be connected to two points in a circuit without having to disconnect any parts or wires or even, if care is taken, to switch the circuit off. A very small current does actually flow through the voltmeter, but with a careful choice of instrument this will only marginally affect the voltages to be measured (section 6.5).

We talk about a voltmeter indicating the p.d. *across* a component or the voltage *at* one point relative to another.

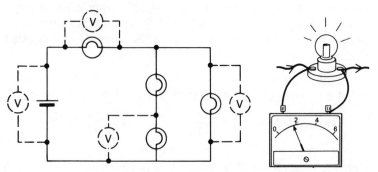

Fig. 3.1   Placing a voltmeter in a circuit

3.9 Voltages or p.d.s in series and parallel arrangements

From what has been said about the idea of electrical 'level' or force, which we call voltage or potential difference, two straightforward consequences follow which will be helpful when considering other circuits later in the book (see below).

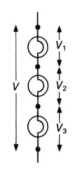

The separate p.d.s V_1, V_2 and V_3 across components connected in series must add up to the total p.d. V across them (like differences of height on a map).

$$V = V_1 + V_2 + V_3$$

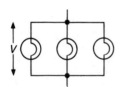

The p.d. V across each component connected in parallel must be the same (the same height difference measured by different routes).

Notice the use of the italic symbol V for a voltage or p.d. on a circuit diagram; V is used as an abbreviation for volt and to indicate a voltmeter.

3.10 Summary

A current is caused by an electromotive force (e.m.f.) or a potential difference (p.d.).

Both e.m.f. and p.d. are measured in volts.

A volt is a joule per coulomb.

The earth is used as the zero of potential.

The size of an e.m.f. or p.d. is indicated by a voltmeter, which is connected in parallel across the component concerned.

In a series connection the p.d.s. across each component may be different.

In a parallel connection the p.d. across each component is the same.

4

Ohm's Law and Resistance

4.1 The idea of a law in science

One of the main objectives of scientists is to try to establish relationships between different quantities. They do this by observing how objects or materials behave towards one another and by making measurements of the changes in one quantity caused by changes in another. If a pattern begins to emerge, the next step is to suggest what kinds of relationships might exist and to set up deliberate experiments to test their validity. Subsequently, if many different experiments fit in with the suggested ideas, the established relationship may be called a *law* and be given the name of one of the original workers in the area of science concerned. A law is thus a summarising statement of what has repeatedly been found to occur, and is usually expressed in terms of the relationship between two quantities under specified conditions.

Work in many branches of science has led to the formulation of laws – for example: *Boyle's Law* about pressure and volume of a gas, *Newton's Laws* dealing with the effects of forces on objects, *Graham's Law* linking the density of a gas with its rate of diffusion, *Mendel's Law* of genetic inheritance, *Faraday's Laws* describing chemical changes in liquids which carry electric currents. The usefulness of having laws lies in the convenient way they can be made to describe not only the *quality* of a relationship (such as that the pressure of a gas rises when it is heated) but also the *quantitative* nature of it (that the pressure increases by $\frac{1}{273}$ of its value at $0°$ C for every degree rise in temperature).

4.2 Ohm's Law – current and p.d.

The first law we met in electricity was Kirchhoff's (section 2.3), which stated simply that the total current entering a junction is the same as that leaving it. The second law we need to use is Ohm's, which concerns the current in a conductor as a result of the potential difference (p.d. or voltage) applied across it.

The simple circuit arrangement of Figure 4.1 can be used to investigate how the current in the conductor X depends on the applied voltage.

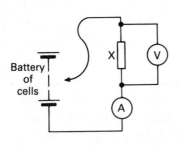

Different voltages, registered by the voltmeter V, are obtained by connecting to different cells in the battery, and the ammeter A indicates the size of the current. (Remember the points mentioned in sections 1.8 and 3.8 about the characteristics of ammeters and voltmeters. Chapters 6 and 9 will return to this matter.)

Fig. 4.1 Measuring current and p.d.

For a conductor such as a piece of iron wire, some typical results might be:

p.d. (in volts)	0	1.5	3.0	4.5	6.0	7.5	9.0
Current (in amps)	0	0.19	0.38	0.56	0.75	0.94	1.05

These results can be displayed in the form of a graph (Figure 4.2).

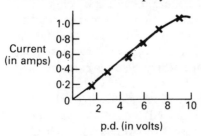

The obvious characteristic of the graph is that, apart from the last point, the pairs of readings lie close to a straight line through (0,0), the *origin* of the graph.

Fig. 4.2 Current – p.d. graph

Similar behaviour is demonstrated by many different conductors, including the departure from the straight line at the high current end. We shall see later that conductors become hotter as the current in them rises (Chapter 8), and that this alters their electrical properties (Chapter 5). It appears, therefore, that there is a simple connection between current and p.d., provided the conductor does not become hot. This is described in *Ohm's Law*:

> For a conductor maintained at constant temperature the current flowing through it is directly proportional to the p.d. across its ends.

In this statement, *directly proportional* means that when the p.d. is doubled the current is also doubled, when the p.d. is trebled the current is trebled, and so on. It also implies both a graph which is a straight line passing through the origin, as in Figure 4.2, and a constant ratio between current and voltage.

4.3 Current and voltage graphs

Strictly speaking, it is only those conductors which conduct an electric current according to the simple graph of Figure 4.2 which can be described as obeying Ohm's Law. Most commonly they are metallic conductors, i.e. iron or copper or alloys such as manganin. Some typical graphs are shown in Figure 4.3 which indicate a variety of conduction curves for different objects or devices.

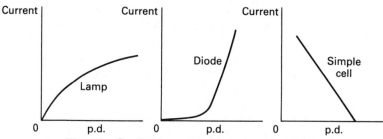

Fig. 4.3 Conduction graphs for various devices

4.4 Limitations on Ohm's Law

There are many devices which clearly do not obey Ohm's Law. In fact it probably works only for metals, and then only under specific conditions (e.g. constant temperature). Nevertheless it is a very useful law which can describe many real situations in a useful way. A wide variety of materials and components do follow Ohm's Law over at least part of their range.

Limitations are found to apply, in practice, to many laws of physics, so Ohm's Law is not unique in this. Often advances in science have been made when a law which was thought to be true was seen not to apply above a certain point, or in a particular range, or under different conditions. Elastic materials, for example, obey Hooke's Law until they are stretched too far, gases fail to obey Boyle's Law under extreme conditions, and a filament lamp obeys Ohm's Law until it begins to get hot. Such limitations do not invalidate the laws themselves, they merely reflect the circumstances under which the laws were discovered or formulated, and define the conditions in which the laws *are* valid and of practical value.

4.5   The idea of resistance

Just as the ordinary word current has been taken into use to describe electrical flow through conductors, so the word *resistance* is used with almost its normal meaning of hindrance or opposition in electrical components. The basic idea of there being a movement of something around a circuit in electricity can be compared with movements of water or traffic or people (section 1.3), and it is a common experience that pipes or roads or corridors can hinder the flow to varying extents. Water will pass more easily through wide pipes than through narrow ones, traffic more easily along motorways than country lanes, people more easily through wide thoroughfares than along narrow corridors. Alternatively, we often talk about the *conductivity* of copper being greater than that of wood or plastics for both electricity and heat flow, describing the ability to allow movement rather than hinder it.

It is as if the 'awkward' conductors of water, traffic or people exert a *resistance* to the flow; those which allow it easily could be said to have a low resistance and those which hinder the flow a high resistance. Deciding which parts of a road should be widened or what sizes of pipes are best in a central heating system, or the width of fire exits in public buildings can all be thought of as problems to do with resistance.

4.6　Electrical resistance – the ohm

For any conductor of electricity, the ratio of potential difference to current flowing is called its *resistance*.

$$\text{Resistance} = \frac{\text{Potential difference}}{\text{Current flowing}}$$

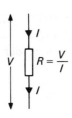

The letter R is used for resistance, and it is measured in *ohms* (named after the German scientist Georg Simon Ohm, died 1854), for which the symbol is Ω (capital omega, the last letter in the Greek alphabet). Since we already have the volt for p.d. and the amp for current, it follows that:

$$1\text{ ohm} \equiv \frac{1\text{ volt}}{1\text{ amp}} \quad \text{or} \quad 1\text{ volt/amp}$$

In words, a conductor has a resistance of one ohm if a p.d. of one volt causes a current of 1 amp to pass through it.

(Notice the symbol ≡ in the above statement. It is not an equation where the left hand and right hand sides are merely equal to each other; they are the same thing. A volt per amp *is* an ohm because that's what we have agreed to call it – not a matter of complicated science, merely of giving a single name to that particular ratio, just as we have seen with the ampere and the volt.)

4.7　Resistances of common devices

As with current and voltage, there is a wide range of resistances met in everyday use. The usual multiple and submultiple prefixes are added to the Ω symbol to show large or small values. Low resistances are not commonly met in domestic appliances so that mΩ and $\mu\Omega$ are quite rare, but kΩ and MΩ are frequently found, especially in radio, TV and other electronic devices. Here are some typical values:

Torch bulb	5 Ω	Pencil lead	20 Ω
100 W lamp	576 Ω	Ammeter	0.01 Ω
Electric Iron element	80 Ω	Voltmeter	10 kΩ
Immersion heater	19 Ω	Loudspeaker	8 Ω
Diode forward	10 Ω	Resistor	almost any value
reverse	50 MΩ		from 0.1 Ω to 100 MΩ

4.8 Resistance calculations

With such a simple relationship between current, voltage and resistance, it is a straightforward task to work out any one of the three if the other two are known. From the basic definition of resistance (section 4.6) we can write:

$$R = \frac{V}{I} \quad \text{or} \quad V = I \times R \quad \text{or} \quad I = \frac{V}{R}$$

and use the appropriate one for the calculation.

For example, if $V = 100$ V and $I = 4$ A, using $R = \dfrac{V}{I}$ we get

$$R = \frac{100}{4} = \mathbf{25\ \Omega}$$

Alternatively, if $I = 3$ A and $R = 5$ Ω, using $V = I \times R$ we get

$$V = 3 \times 5 = \mathbf{15\ V}$$

Similarly, if $V = 12$ V and $R = 3$ Ω, using $I = \dfrac{V}{R}$ we get

$$I = \frac{12}{3} = \mathbf{4\ A}$$

Care is needed when the quantities are expressed in multiples or submultiples (e.g. kV, mA or MΩ), because unless the noughts are included in the calculation the result will be incorrect.

For example, if $V = 5$ kV and $R = 2$ MΩ, using $I = \dfrac{V}{R}$ we have

$$I = \frac{5 \times 1000}{2 \times 1\,000\,000} \quad = \frac{2.5}{1000}\,\text{A} = \mathbf{2.5\ mA}$$

The reader might like to work out the missing quantity in each line of the following table:

Current, I	Potential difference, V	Resistance, R
2 A	6 V	3 Ω
10 A	240 V	24 Ω
3 A	18 V	6 Ω
0.1 A	15	150 Ω
4 A	12 V	3 Ω
4 A	220 V	55 Ω
12 mA	120 V	10 K Ω
2 mA	50 K	100 Ω
.2 m A	5 kV	100 Ω
75 mA	150 kV	2 Ω
100 mA	33 V	33 kΩ
4 A	2 kV	0.5 MΩ

4.9 Resistors

It is often necessary for currents and voltages to be of particular sizes to operate devices in electrical circuits, and this is most commonly achieved by using *resistors* manufactured to give appropriate values of resistance. A resistor is a circuit element made of thin wire wrapped on an insulating former, or of a carbon and clay mixture moulded into a cylindrical shape, or of a thin metal oxide film. Connections are made through wires to each end of the resistor (Figure 4.4).

For high precision work the relatively expensive wirewound type has to be used; the metal oxide ones can be made more cheaply but only to less precise values. Carbon resistors are the most commonly used for applications where high accuracy is not required.

The circuit symbol for a resistor ─▭─ was met in section 4.2; one whose resistance can be varied (e.g. the volume control on a radio) is indicated by ─▱─ .

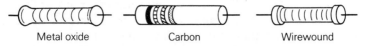

Metal oxide   Carbon   Wirewound

Fig. 4.4   Types of resistor

4.10   Preferred values for resistors

For many purposes the value of a fixed resistor does not need to be precise, a tolerance of anything up to ± twenty per cent of the stated value being sufficiently accurate. It is therefore not necessary for manufacturers to produce more than a certain range of values, provided the gaps between them are within twenty per cent either way. Starting with 10 Ω, the following series of values is all that is needed up to 100 Ω if each one could be up to twenty per cent in error either way:

$$10 \quad 15 \quad 22 \quad 33 \quad 47 \quad 68 \quad 100$$

(e.g. $15 + 20\% = 18$ and $22 - 20\% = 17.6$ so that the values overlap to twenty per cent accuracy).

Similarly, for a tolerance of ± ten per cent, a more closely spaced series would be needed (which includes the twenty per cent series):

$$10 \quad 12 \quad 15 \quad 18 \quad 22 \quad 27 \quad 33 \quad 39 \quad 47 \quad 56 \quad 68 \quad 82 \quad 100$$

These sizes are known as the *preferred values* for resistors for ten per cent and twenty per cent tolerances. It would be possible also to have smaller divisions for a five per cent series.

With multiples and submultiples of the preferred values, all necessary ranges can be covered, e.g. 330 Ω, 4.7 kΩ, 2.2 MΩ.

4.11   Colour code for resistors

The nominal resistance of a fixed resistor is sometimes printed on it in figures, e.g. 6.8 k for 6.8 kΩ, but more often coloured rings are used, either three or four of them according to the following code:

Black	0	Green	5	
Brown	1	Blue	6	Silver ± 10%
Red	2	Violet	7	Gold ± 5%
Orange	3	Grey	8	
Yellow	4	White	9	

Three colours are used to indicate the value of a resistor: the first and second show the figures in the preferred series, the third indicating the number of noughts after them. The fourth colour, silver or gold, shows the manufactured tolerance of the resistor at

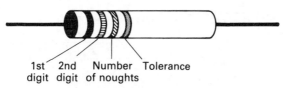

1st 2nd Number Tolerance
digit digit of noughts

Fig. 4.5 Resistor coding

ten per cent or five per cent (Figure 4.5). If there is no fourth ring, the tolerance is twenty per cent.

Thus a sequence of brown, green, red (1, 5, 2), silver, means a resistor of 1500 Ω at ten per cent tolerance, or 1.5 k̄Ω. Similarly yellow, violet, orange implies 47 000 or 47 kΩ at twenty per cent tolerance. The reader might work out the values indicated by the following colour sequences:

blue	grey	brown	gold	680Ω 5%
orange	white	red	silver	3.9KΩ 10%
red	red	green		2.2MΩ 20%
brown	black	yellow		1MΩ 20%
green	blue	black	gold	56Ω 5%
grey	red	orange	silver	82KΩ 10%
yellow	violet	green		4.7MΩ 20%
orange	orange	orange		33KΩ 20%

4.12   Resistance and Ohm's Law

The simple graph of current I against voltage V for a conductor (Figure 4.2) is a straight line showing direct proportion between the two quantities (except at higher values when the conductor gets hot). Such a line implies a relationship between I and V of the kind $I = gV$ or $hI = V$ where g and h are fixed quantities. A set of lines such as those in Figure 4.6 illustrates this point clearly for $g = \frac{1}{2}$, 1, 2, 3, but for any straight line which passes through (0,0) there will be a value for g which could match it.

Thus the equation $IR = V$ will fit such a line and the size of R will

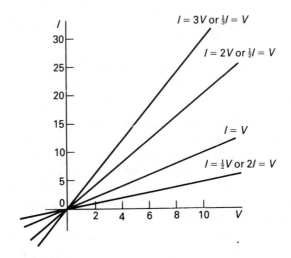

Fig. 4.6   Graphs of $I = gV$ or $hI = V$

relate to the steepness of it: a high R gives a gentle slope and a low R a steep one. So, remembering how resistance was defined in section 4.5, it is clear that on a current–voltage graph the steepness of the line indicates the resistance of the conductor.

A straight line through (0,0) has the same slope at all places along it, so a conductor which obeys Ohm's Law (i.e. for which the I-V graph is this type of line) will have a constant resistance, the same at all points. A piece of wire could have a resistance of 5 Ω, say, if the current were anywhere in the range over which its graph was straight.

We know, however, that many devices do not follow Ohm's Law and their I-V graphs are not straight, as in Figure 4.6. How can the idea of a resistance be applied to those devices? The two ways most frequently used involve the steepness of the line, even if it is not straight, and a spot value at a particular place on the line. In both cases the resistance of the device relates only to a section or a particular point on the graph and is not therefore a constant value. As an example, consider the current–voltage graph for an electric filament lamp (Figure 4.7).

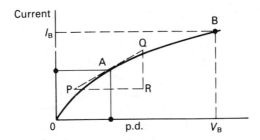

Fig. 4.7 Two measures of resistance for a lamp

At the point A, the steepness of the curve can be taken as that of the dotted line PQ (called a tangent to the curve at A) and used as a measure of the resistance over a range of currents and p.d.s around A. The ratio of the voltage represented by PR to the current represented by RQ gives the size of the resistance. Alternatively, the resistance at point B could be regarded as the ratio of the actual voltage and current operating at that point, i.e. V_B/I_B.

Both ways of looking at the resistance give values in ohms (if V and I are in volts and amps) and each offers a useful working value for resistance. If, for instance, the device were to be used over a changing range of voltages and currents around point A, the gradient method would be the best way to describe its resistance. On the other hand, if the lamp were being run predominantly at B as its operating point, the spot value would give a better guide to its resistance at that particular current and p.d.

The point to stress is that the idea of resistance for a device as the ratio of voltage to current can be used even if the device does not obey Ohm's Law. The equation $V = IR$ is still applicable over a range of operating conditions even if the I-V graph is not a straight line and direct proportionality does not exist.

4.13 Resistance and temperature

Nearly all conductors suffer a change of resistance when they become hot. The commonest example is the filament of an ordinary electric lamp or torch bulb which becomes white hot in normal use. Its resistance is much greater at the high temperature than when it is cold. Most metallic conductors show this pattern of behaviour. When cooled to extremely low temperatures (near −273°C) it is even possible for the resistance to become virtually zero. When this happens the materials are called *superconductors* and can carry an electric current without any heat being developed.

Semiconductors such as silicon or germanium, on the other hand, are affected in the opposite way – their resistance falls if they are heated, and by much larger amounts than metals. Such a property can be useful in electronic fire alarms, for instance, but it can also cause a diode or transistor to fail – if the diode conducts more current which causes it to heat up, which enables it to conduct even more, which results in further heating, etc. This is called *thermal runaway* and has to be avoided in the design of electronic equipment.

4.14   Resistors in series

In a series connection the same current flows through each resistor (section 2.10) but the potential differences across them can be different (section 3.9). The effective or equivalent resistance R of the separate resistors R_1, R_2 and R_3 in Figure 4.8 can be worked out from the overall current and voltage values.

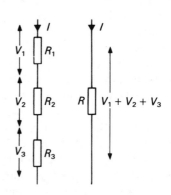

Fig. 4.8   Equivalent resistance in series connection

Since the overall p.d. V must equal the sum of the separate p.d.s V_1, V_2, V_3 we can use the relation $V = IR$ for each of them:

$$V = V_1 + V_2 + V_3$$

so that

$$IR = IR_1 + IR_2 + IR_3$$

or

$$R = R_1 + R_2 + R_3$$

The equivalent resistance for a set of resistors in series is therefore simply the sum of the individual resistances. For example, for three resistors of 4, 6 and 12 Ω in series, the equivalent resistance is 4 + 6 + 12 = 22 Ω. The equivalent resistance in series is always *greater* than that of any of the separate ones. (Water flowing through a series of narrow tubes would be hindered by each of them along the way.)

4.15 Resistors in parallel

When connected in parallel (section 3.9) it is the p.d. that is the same for each resistor and the currents that can be different (section 2.10). Again, the equivalent resistance R can be calculated from the current and p.d. values (Figure 4.9).

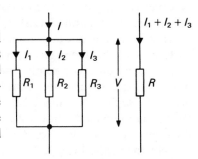

Fig. 4.9 Equivalent resistance in parallel connection

$$I = I_1 + I_2 + I_3$$

so that

$$\frac{V}{R} = \frac{V}{R_1} + \frac{V}{R_2} + \frac{V}{R_3}$$

or

$$\frac{1}{R} = \frac{1}{R_1} + \frac{1}{R_2} + \frac{1}{R_3}$$

The equivalent resistance for a set of resistors in parallel has to be worked out from this equation. For example, for three resistors of 4, 6 and 12 Ω in parallel, the equivalent resistance is worked out by:

$$\frac{1}{R} = \frac{1}{4} + \frac{1}{6} + \frac{1}{12} = \frac{3 + 2 + 1}{12} = \frac{6}{12} = \frac{1}{2}$$

giving $R = 2\ \Omega$

In this case the equivalent resistance is always *less* than that of any individual resistor. (Water flowing through a set of tubes in parallel would have alternative paths available and so be hindered less by the combination than by any single tube.)

If only two resistors are involved in a parallel connection, the equation can be written as their product divided by their sum:

$$R = \frac{R_1 R_2}{R_1 + R_2}$$

Alternatively, a graphical method can be used to find the equivalent resistance for two parallel resistors (Figure 4.10). In this method the horizontal scale is not important. The vertical scale is chosen for each working and the values of R_1 and R_2 marked off at A and B, one on each side of the graph. Lines are then drawn from A to the bottom of the R_2 scale and from B to the bottom of the R_1 scale. Where these lines AQ and BP cross indicates the equivalent resistance R on the same vertical scale. (Readers might like to convince themselves that the distance PQ really does not matter and to find out why the method works.)

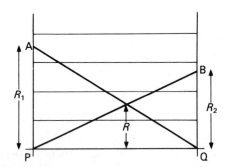

Fig. 4.10   Graphical method for two resistors in parallel

4.16   Networks of resistors

Circuits which contain mixtures of series and parallel connections can be simplified to a single equivalent resistance by taking small sections one at a time. Some examples are illustrated in Figure 4.11.

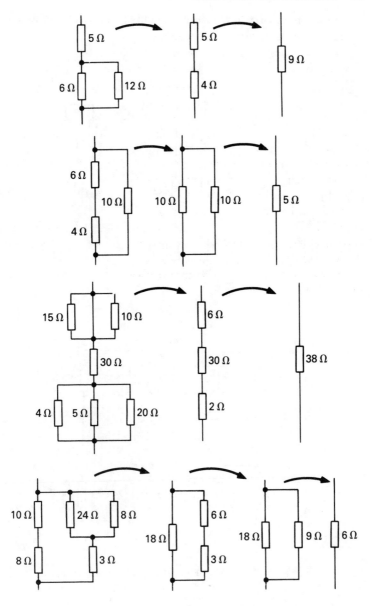

Fig. 4.11 Resistor networks

Readers might like to work out the equivalent resistances for each of the following networks:

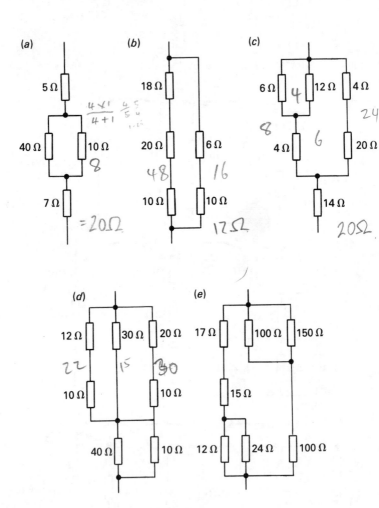

Fig. 4.12

4.17 Resistance in a.c. circuits

As will become clear in Chapter 17, when alternating currents are used there are other factors apart from resistance which hinder the flow of charge: they are called *capacitance* and *inductance*, and all three effects contribute to what is known as the *impedance* offered to the flow of charge. It is even possible to have a negative resistance. Looking back to sections 4.3 and 4.13, the reader might recognise the circumstances when such an idea could be useful.

4.18 Summary

A Law describes the relationship, established by experimental measurements, between two quantities under specified conditions.

Ohm's Law states that the current passing through a conductor is directly proportional to the p.d. across it, provided the temperature remains constant.

Many electrical components do not obey Ohm's Law.

Electrical resistance hinders the flow of charge through a conductor.

Resistance is the ratio of p.d. to current and is measured in ohms.

An ohm is a volt per amp.

$$R = \frac{V}{I} \quad V = IR \quad I = \frac{V}{R}$$

Resistors are made to preferred values according to their precision, their sizes being indicated by a colour code.

Direct proportion implies a straight line graph through the origin.

Resistance changes with temperature.

For resistors in series, $R = R_1 + R_2 + R_3$

For resistors in parallel, $\frac{1}{R} = \frac{1}{R_1} + \frac{1}{R_2} + \frac{1}{R_3}$

5

Sources of Electricity

5.1 Primary cells

It was discovered as long ago as 1790 by Volta that when two dissimilar conducting materials were in contact with a conducting liquid, a voltage was developed between them. Copper and zinc in sulphuric acid formed a simple 'wet' *primary* cell, and a more elaborate version developed by Daniell was widely used for laboratory work (Figure 5.1).

It was soon recognised that other conducting liquids or pastes could be used as *electrolytes*, as could different metals such as nickel and cadmium, and even carbon as *electrodes*. The Leclanché cell was used in both wet and dry forms, the latter being common still in torches, toys, transistor radios, electric bells, etc. (Figure 5.2).

Several cells of a similar type but using different chemicals have

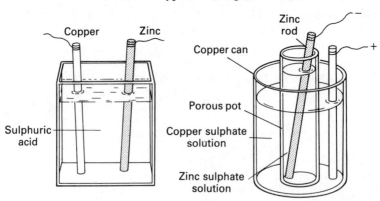

Fig. 5.1

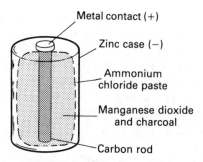

Metal contact (+)

Zinc case (−)

Ammonium chloride paste

Manganese dioxide and charcoal

Carbon rod

Fig. 5.2   Dry Leclanché cell

also been developed, some being able to deliver steady currents for surprisingly long periods of time. Some cells can be made in very small sizes and find uses in such things as electronic watches, hearing aids and cardiac pacemakers.

5.2   Secondary cells (accumulators)

Whereas primary cells generate electricity from chemical action between the substances inside them, *secondary* cells have to be charged up by other means before they can produce current of their own. The most common example is the car battery, a lead-acid accumulator, which is normally kept charged by the action of a generator in the car (Figure 5.3).

A car battery is usually six cells connected in series. When charged, one set of plates becomes coated with lead peroxide (lead

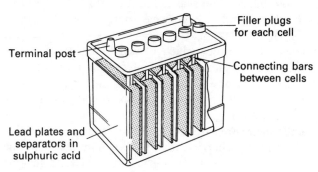

Filler plugs for each cell

Terminal post

Connecting bars between cells

Lead plates and separators in sulphuric acid

Fig. 5.3   Lead-acid accumulator

(IV) oxide) which, with other lead plates and the sulphuric acid, generate the electrical voltage. As the battery discharges in use, the concentration of the acid changes due to the production of water, and its level has sometimes to be topped up with distilled or ion-free water. The chemical reactions are quite complex, but because they are reversible this type of cell can be repeatedly charged up and reused. Electric vehicles such as milk floats or fork lift trucks use accumulators of this kind and have to be recharged, often at night, for each day's use.

Smaller rechargeable cells, with electrolytes in paste form, have also been developed for small domestic appliances such as shavers or torches, using different materials for the electrodes and electrolytes.

5.3   Other sources of electricity

Solar cells convert light directly into electrical energy. They are used in calculators and as power supplies in space probes, amongst other things.

Piezo-electric crystals such as quartz produce electricity when they are twisted. Gas igniters, hi-fi pickups for record players, and cigarette lighters often employ this effect.

Fuel cells can 'burn' fuel in such a way that it is converted into electricity directly. Hydrogen-oxygen cells have been used in space craft where the by-product, water, is also vitally needed.

Dynamos and generators develop electricity from rotating machinery such as turbines or internal combustion engines. They are used in motor vehicles and in large scale energy production in power stations (Chapters 11 and 13).

Power packs are devices which convert a particular electricity supply such as the mains or a car battery into a more convenient form for specific purposes, e.g. driving an amplifier or other electronic equipment where precise currents or voltages are required (Chapters 11 and 12).

Electrostatic generators can be made to produce extremely high voltages by using the charges developed when suitable materials are rubbed or rolled on one another. The Van der Graaff generator is one example (Chapter 14).

5.4 Electromotive force (e.m.f.)

Whatever the source of electricity used for a particular purpose, the main objective is to have something which can move electrons through a circuit. What is wanted is a *force to move electrons*, usually called *electromotive force* or *e.m.f.*, and it is the task of the source of electricity to produce such a force. Like potential difference, e.m.f. is measured in volts. The size of the e.m.f. developed by any device depends only on how it is made. Each type of source has its own characteristic e.m.f.

In section 3.3 a volt was defined as a joule per coulomb. What this means is that if a coulomb of charge is moved from one place to another and a joule of work is done in the process, the p.d. between the two places is said to be one volt. If two joules of work are done, the p.d. is two volts (Figure 5.4).

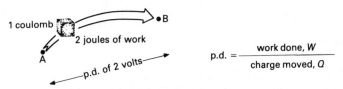

$$\text{p.d.} = \frac{\text{work done, } W}{\text{charge moved, } Q}$$

Fig. 5.4 Potential difference between two points

In the case of a complete circuit, however, charges are moved continuously around the loop, through the source of electricity itself and whatever else is in the circuit. It takes energy to do this, of course, which the source supplies. The e.m.f. of the source is defined as the work the source does for every coulomb which flows completely round the circuit (Figure 5.5).

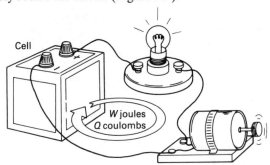

Fig. 5.5 E.m.f. of a cell = W/Q for complete circuit

The e.m.f. is the maximum voltage the source can produce (though as we shall see in section 5.5 it is often not all available for external use). Here are some typical values for e.m.f.s:

Hi-fi pickup	1 mV	Dry battery	9 V
Simple cell	1 V	Car battery	12 V
Fuel cell	1 V	Piezo-electric unit	2 kV
Daniell cell	1.1 V	Power station	4.4 kV
Dry (Leclanché) cell	1.5 V	Van der Graaff	1 MV
Cycle dynamo	6 V	Power pack	1 V to 1 kV

5.5   Internal resistance

Although the e.m.f. generated by a source is of great importance for any application, of equal significance is the current it can deliver. Since the e.m.f. is decided by the make-up of the source, the current will depend on the resistance in the circuit. The new point here is that the source itself might have some resistance, and since the current must pass round the complete circuit including the source, this will have to be taken into account as part of the total resistance.

A simple demonstration illustrates the presence of internal resistance in cells. If a cell is short-circuited, i.e. its terminals connected directly by a thick copper wire, the effects vary surprisingly (Figure 5.6). In the case of a 2 V lead-acid accumulator cell, the copper wire will glow red and any insulation on it will probably burn off. (*Warning*: this does not do the cell any good!) The dry battery,

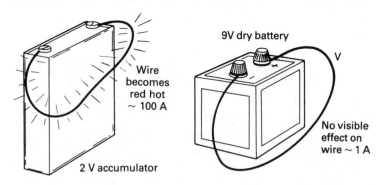

Fig. 5.6   Internal resistance reduces current

although of higher e.m.f. than the accumulator, produces no visible effects on the wire.

If an ammeter were included in each circuit it would indicate perhaps 100 A in one case but only 1 A in the other. Why is the current so much greater from the accumulator than from the dry battery? The only explanation is that the resistance in the circuit is very much lower, i.e. the lead-acid cell has a very low *internal resistance* (say $1/50 \, \Omega$) compared to the dry battery (say $10 \, \Omega$).

The size of the internal resistance of a source can be seen to be vital in determining how much current it is capable of delivering. Clearly, the choice of source can be affected as much by the current required from it as by its e.m.f. Convenience and safety are also factors in any particular case, of course, but there is little option to using accumulators in motor cars; dry batteries are often used for electronic devices though power packs are generally more economical in practice. A significant internal resistance is a useful feature to avoid damage if the source is accidentally short-circuited. Generally the larger dry cells have lower internal resistance than smaller ones for the same voltages.

5.6 Terminal p.d. and e.m.f. – Ohm's Law for complete circuits

Any source of electricity presents two characteristics to the user – e.m.f. (E) and internal resistance (r), both of which are in effect located within its terminals.

The source can deliver a current (I) to an external resistance (R). The p.d. (V) between the source's terminals is what the external circuit 'sees' (Figure 5.7). The source, however, has its e.m.f. with

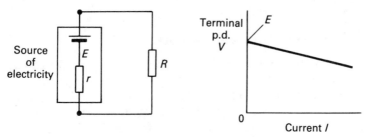

Fig. 5.7 Terminal p.d. and e.m.f.

which to send the current through both internal and external resistors and has to supply the work for each coulomb which passes through both R and r. Its e.m.f. can therefore be written as the sum of the two voltages:

$$\boxed{E = IR + Ir}$$

For the external resistance $V = IR$, so we get

$$E = V + Ir$$

where V is the p.d. across R (and across the cell's terminals), or:

$$\boxed{V = E - Ir}$$

These two highlighted equations represent the relationship for the whole circuit and give the two measurable quantities I and V in terms of the source's properties E and r. The most obvious feature of the second equation is that V is always less than E when the source is delivering a current I, as long as there is an internal resistance r. If r remains constant (which is approximately true for dry cells and also for power packs) the relationship between V and I is shown in the graph, with V equalling E when the cell is delivering no current (Figure 5.7).

If a source has no (or negligible) internal resistance, i.e. r is effectively zero, then $V = E$ for all values of I. Accumulators have very small internal resistances and can therefore be regarded as maintaining fixed p.d.s for external use, virtually the same as their e.m.f.s. Care must be taken, though, to avoid dangerously large currents arising from short-circuiting.

5.7 Cells in series and parallel

Series Cells are often packed together in series to make batteries (e.g. six 1.5 V dry cells to form a 9 V battery) or used separately in series one behind the other in torches or transistor radios to provide the desired voltage. As would be expected, the result is an e.m.f. and an internal resistance equal to the sums of those of the separate cells (Figure 5.8).

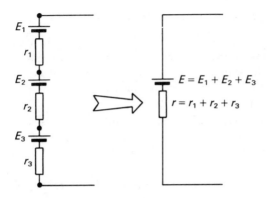

Fig. 5.8 Cells in series

As both e.m.f. and internal resistance are raised, the current does increase but not in proportion to the number of cells. (The reader might care to work out a simple case using $E = IR + Ir$ for $E = 2$ V, $r = 2\ \Omega$, $R = 5\ \Omega$, then again with three cells giving $E = 6$ V, $r = 6\ \Omega$, and $R = 5\ \Omega$ still.) The main advantage of using cells in series is to match the internal resistance to the external resistance; when this occurs, the latter has maximum *power* available to it (Chapter 8).

Parallel Only cells of equal e.m.f.s would ever be joined in parallel, since there would be no point in the cells sending current through one another. In this case the e.m.f. of the combination remains unchanged but the internal resistance is reduced as indicated in Figure 5.9.

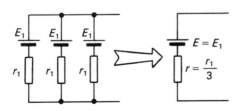

Fig. 5.9 Cells in parallel

5.8   Summary

Primary cells produce electrical energy from their own chemical reactions, whereas secondary cells need to be charged up first.

Other sources of current include solar cells, piezoelectric crystals, fuel cells, generators, power packs, thermoelectric and electrostatic generators.

Sources of electricity have e.m.f.s and internal resistances.

For a complete circuit, $E = IR + Ir$ and $V = E - Ir$.

For cells in series, e.m.f. $= E_1 + E_2 + E_3$, internal resistance $= r_1 + r_2 + r_3$.

For n similar cells in parallel, e.m.f. $= E$, internal resistance $= \dfrac{r}{n}$.

6

Measuring Electricity

6.1 Ammeters and voltmeters

Earlier chapters have introduced the idea of meters being used to measure and indicate currents and potential differences – ammeters and voltmeters respectively. Such instruments are reliable and easy to use, but some care is needed in practice to ensure that they are not damaged by misuse.

Sensitivity   The most important point is to choose the instrument best suited to the purpose in hand, especially as regards the sensitivity required. An ammeter, for instance, will have a limit to the current it can handle, i.e. that current which takes the pointer to the end of its scale. This value is normally clearly marked on the scale: currents larger than the maximum can damage the moving parts inside the instrument or at least bend the pointer.

The same consideration applies to the selection of a voltmeter – look for its maximum reading and choose one that can cope with the p.d. you wish to measure. Figure 6.1 shows some typical ammeter and voltmeter ranges.

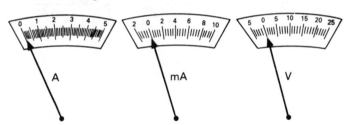

Fig. 6.1   Typical ammeter and voltmeter ranges

+ and − connections   Some meters have an offset zero mark which enables them to indicate a small range of negative values, i.e. currents or voltages in the reverse direction from that expected. The conventional current direction should be connected into the positive terminal of the meter, as should the high voltage level. If the connections are the wrong way round the pointer will move to the left instead of the right and the wires should be reversed. The positive terminal of a meter is often coloured red or marked with a +.

Scale readings   Figure 6.1 also shows that the size of the scale divisions can vary depending on the range of the instrument. Careful observation is needed to be sure that the position of the pointer is read accurately between the numbered markings on the scale.

a.c. and d.c.   Basic meters are designed to be used on d.c. *or* a.c. but not on both. The mark ═ indicates d.c. whilst ≈ indicates a.c. A d.c. meter will not operate at all with alternating currents. Some a.c. meters will respond to direct currents but would not indicate the correct values.

6.2   Resistances of ammeters and voltmeters

When a meter such as an ammeter or voltmeter is connected into a circuit it is very important that the instrument itself does not affect the size of the current or p.d. it is intended to measure. To illustrate the point for an ammeter, Figure 6.2 shows a simple circuit and then the effects of using two different ammeters to measure the current.

The *low* resistance ammeter has only a very small effect on the current.

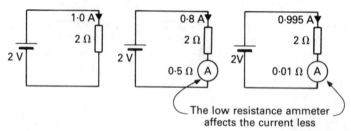

Fig. 6.2   Effect of an ammeter on the current to be measured

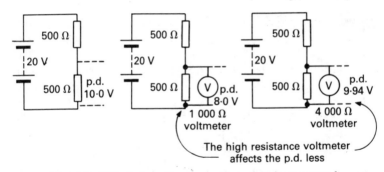

Fig. 6.3 Effect of a voltmeter on the p.d. to be measured

Figure 6.3 illustrates the corresponding case with a voltmeter.

This time it is the *high* resistance voltmeter which has little effect on the p.d.

Ideally, an ammeter should have a zero resistance and a voltmeter an infinitely high one, but neither of these limits is easy to attain. What matters is that the resistances should be high or low relative to those of the circuit into which the meter is to be connected. As a rough rule of thumb, a factor of at least twenty times greater or smaller should be allowed and the instrument chosen with that in mind, as well as its having the appropriate range, of course.

6.3 Galvanometers

The basis for almost all ammeters and voltmeters is an instrument called a *galvanometer* (named after the Italian experimenter, Luigi Aloisio Galvani d. 1798). This is a sensitive device for detecting currents which uses the twisting of a rectangular coil of wire, through which the current flows, between the poles of a strong magnet (see Chapter 9). Figure 6.4 shows one such instrument. The specification on the meter gives its scale range, the unit of current (or voltage) and its resistance. Galvanometers can be manufactured with a variety of current range and resistance, e.g. 10 Ω 10 mA, or 100 Ω 100 μA, or 50 Ω 1 mA, and so on.

The resistance of a galvanometer, which is mostly in the fine copper wire on its coil, is commonly between 10 Ω and 100 Ω, which is neither small enough nor large enough for it to be used directly as

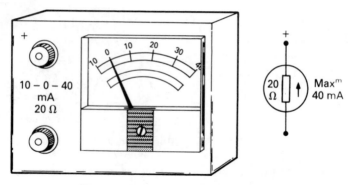

Fig. 6.4 Basic galvanometer and symbol

an ammeter or voltmeter. It is possible, though, for a galvanometer to be converted into either by quite simple means.

6.4 Conversion to ammeter

For a galvanometer with a range of only 40 mA to be adapted to read up to four amps, say, most of the current must be diverted from the meter itself through a *shunt*, i.e. a resistor in *parallel* with the meter. Of the four-amp current, all but 40 mA passes through the shunt and the resistor is chosen so that it divides the current in exactly those proportions. A fairly simple calculation shows that the shunt needs to have a resistance of 0.202 Ω. In a similar way the resistance of a shunt can be chosen to give the combined galvanometer and shunt a range up to any required value. A 2.222 Ω shunt would be needed, for example, to convert the same galvanometer to read up to 400 mA. Figure 6.5 shows how a shunt could be added to the basic galvanometer, and also the electrical arrangement implied.

Manufacturers will generally supply a variety of shunts to fit a galvanometer, so that several current ranges can be used. Sometimes, to avoid steps of only 10, 100, 1000 . . . times the basic range of the galvanometer two different scales are provided for the pointer to indicate. In the case of the meter in Figure 6.4 a second scale of 20-0-80, for example, could be marked alongside the 10-0-40 scale (as in Figure 6.10). The user then has to decide which figures match the range shown on the particular shunt used, to be

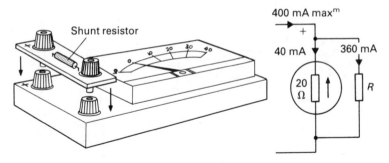

Fig. 6.5   Conversion of a galvanometer to an ammeter

careful about how many noughts are appropriate and to read the subdivisions on the scale accurately.

Using a shunt in parallel with a galvanometer means that the overall resistance will be reduced. For the 20 Ω 40 mA galvanometer described above, the converted instrument has a resistance of 2 Ω with its 400 mA shunt and 0.2 Ω with its 4 A shunt. This low resistance is just what is needed for an ammeter.

6.5   Conversion to voltmeter

A galvanometer with a resistance of 20 Ω capable of carrying only 40 mA could be connected to a p.d. of up to 40 mA × 20 Ω = 800 mV or 0.8 volt (using $V = IR$). Any p.d. in excess of this would give a current larger than the 40 mA maximum. To cope with higher voltages than 0.8 V a resistor must be placed in *series* with the galvanometer, called a *multiplier*, which can limit the current to no more than 40 mA through the combined multiplier and meter. The resistance of the multiplier is chosen according to the range of p.d. to be measured. For example, one of 180 Ω would allow this galvanometer to read up to 8 V, and of 1980 Ω up to 80 V. Figure 6.6 illustrates how a slightly different plug-in unit could be made to convert the galvanometer for use as a voltmeter, and the corresponding circuitry.

As in the case of the ammeter, a variety of multipliers would be available with the range conversion indicated on them. As far as resistance is concerned, the multiplier and galvanometer resistances are in series and simply add together to give 200 Ω for the 8 V

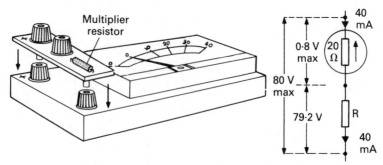

Fig. 6.6 Conversion of a galvanometer to a voltmeter

range and 2000 Ω for the 80 V range – high resistances being desirable for voltmeters.

6.6 Ohmmeters

Galvanometers essentially respond to currents flowing through them. For a passive quantity like resistance to be measured, it has to be arranged for a current to be sent through the resistor and to be indicated by a galvanometer. (A high resistance will result in a low current, however, and a low resistance in a high current, so any resistance scale added to a galvanometer will inevitably read the wrong way round.) Figure 6.7 shows a way of arranging things, the resistance to be measured being connected between X and Y.

If a zero resistance (thick copper wire, say) is joined to X and Y, the galvanometer must be protected from a current greater than its maximum scale reading, e.g. 40 mA, and the resistor R serves this purpose. From the values given in Figure 6.7 the resistance R can be calculated as 55 Ω ($E = I(R+20)$, being careful about units). A zero resistance then allows 40 mA to pass and the galvanometer pointer goes to its full scale deflection, which must therefore be marked 0(nought) on a resistance scale.

Similarly if X and Y are not joined by anything, i.e. infinite resistance, the galvanometer indicates zero current and this position becomes ∞, infinity, on the resistance scale. Intermediate values between 0 and ∞ must be worked out, running from right to left. Readers might find this task not too difficult if they realise that in Figure 6.7 the total resistance of 55 + 20 = 75 Ω allows 40 mA with

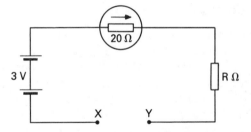

Fig. 6.7 Basic ohmmeter design

X–Y joined together through zero resistance. An extra 75 Ω between X and Y will therefore reduce the current by half, meaning that 75 Ω on the resistance scale will fall under the 20 mA current mark. Other values can be treated similarly until a full set of divisions is developed on the resistance scale. Figure 6.8 shows the result of doing this for the particular circuit shown above. A more sensitive galvanometer would be required for accurate readings above 1000 Ω.

One curious feature of the resistance scale is that it is not evenly divided. It would never be possible to get up to ∞, of course, if it were! In practice there would be a small variable resistor in series with *R*, used as a zero-set control, to allow for changes in the e.m.f. of the cells.

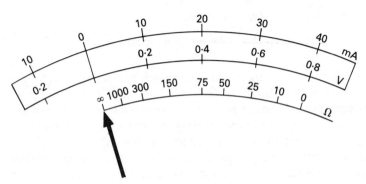

Fig. 6.8 Scales for current, voltage and resistance

6.7   Multimeters

Separate ammeters, voltmeters and ohmmeters with appropriate shunts and multiplier units to provide a variety of ranges are very flexible in use and necessary when measurements of the different quantities are required simultaneously. For many other purposes, though, it is convenient to combine current and voltage ranges, and even resistance ranges too, into a single multi-range instrument called a *multimeter*. A simple design for such a meter is shown in Figure 6.9, in which separate terminals are needed for currents and voltages/resistances. Readers might like to trace the connections at switch positions 2, 4 and 7 to compare with Figures 6.5, 6.6, and 6.7.

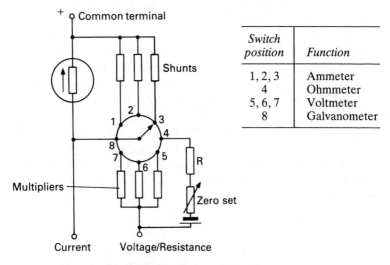

Switch position	Function
1, 2, 3	Ammeter
4	Ohmmeter
5, 6, 7	Voltmeter
8	Galvanometer

Fig. 6.9   The design of a multimeter

It is possible, by a more elaborate switching arrangement, for just two terminals to be used, as in a simple meter. A commercial multimeter might have a range of d.c. amps and volts, a.c. volts and ohms (Figure 6.10). The a.c. voltage ranges are provided through the use of a transformer and rectifier (Chapter 12). It is possible to include a.c. current ranges, but these require more advanced components and circuitry. Note the 'reset' button, an automatic cut-out which is tripped if too large a current is inadvertently sent through the galvanometer movement, which is easy to do with multimeters if the selector switch is left in the wrong position when changing the measurement to be made. It is advisable always to return the selector switch to the OFF position after use or between different uses. Another useful feature on some multimeters is a ÷2 button which doubles the scale reading on current and p.d. ranges when pressed. When the pointer indicates a small reading but one too large to be accommodated on the next range, the accuracy of interpreting the pointer's position is increased with a larger deflection. The user must remember, though, to divide the new scale reading by two!

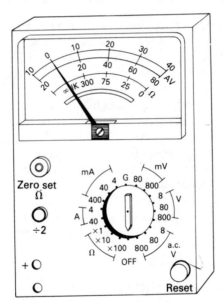

Fig. 6.10 A practical multimeter

6.8 Cathode Ray Oscilloscope (CRO)

One important and very different instrument which can be used in many branches of electricity to measure or display electrical quantities is the *cathode ray oscilloscope*. This device uses a beam of electrons (cathode rays) which produce light when the beam strikes

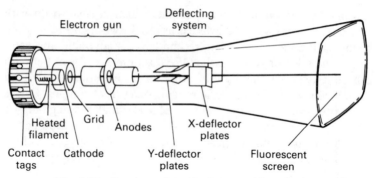

Fig. 6.11   Features of a cathode ray oscilloscope tube

a specially coated screen. Figure 6.11 shows how this is done.

A very high vacuum is required inside the tube, and the electrons, which come from a hot filament, are accelerated down the tube by a p.d. of about 2000 V. A system of cylinders and discs with suitable voltages focuses the beam to a fine spot on the screen, and the number of electrons per second in the beam can also be controlled. Deflector plates, between which the beam travels, can be used to manoeuvre the spot in two directions (X – horizontally, Y – vertically) to any point on the screen. One of each of the X- and Y-plates is connected to earth, and voltages can be applied to the other plates to deflect the beam.

The special virtue of the CRO is that electrons are the only moving part and they can respond immediately to voltage changes on the deflector plates. They travel down the tube at about one-tenth the speed of light.

6.9   Setting up the CRO

The controls of a CRO are often arranged in a similar way to Figure 6.12 and the first task is to make the electron travel down the tube and land on the screen. The procedure is as follows:

(*a*)   Switch on, allow a warming time of about 20 seconds;
(*b*)   turn up the brightness control;
(*c*)   make sure the time base is in the 'OFF' position;
(*d*)   adjust the X- and Y-shift controls to centre the spot of light on the screen;

(*e*) if necessary adjust the focus control to obtain a sharply defined spot (easier at less than maximum brightness).

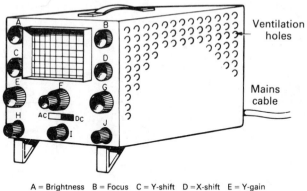

A = Brightness B = Focus C = Y-shift D =X-shift E = Y-gain
F = Time base G = X-gain H = Y-input I = Earth J = X-input

Fig. 6.12 Possible layout of an oscilloscope

Voltages can be applied to X- and Y-plates through the terminals marked, using earth as one connection for each. The a.c./d.c. switch usually operates only on the Y-plates and is useful for distinguishing between the absolute voltage level (d.c.) and fluctuations around a steady value (a.c.). With an a.c. setting, only *changes* of voltage are registered, so that a swing from 110 V to 90 V would appear as one from +10 V to −10 V. The X- and Y-gain controls can allow for wide ranges of input in a series of switched settings of varying sensitivities. X- and Y-shifts can be used to move the spot to any desired place on the screen.

6.10 Effects of d.c. and a.c. inputs

The CRO is a voltage-operated device responding to p.d.s across the X- or Y-plates, taking virtually no current and therefore acting as a perfect voltmeter (section 6.2). Figure 6.13 shows the effects of applying d.c. and a.c. voltages to the CRO (switched to d.c.).

Note that the spot of light is drawn into a line if the frequency of the a.c. is above a few H7 because of its rapid movement across the screen. The amount of movement of the spot away from the centre will depend on the setting of the X- and Y-gain controls, and for a given setting is proportional to the voltage applied. For a.c. input

(*a*) No input

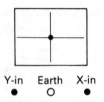

(*b*) Input to X

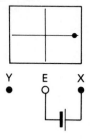
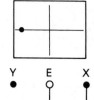
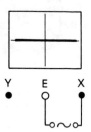

(*c*) Input to Y

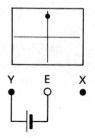

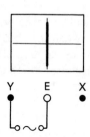

(*d*) Input to X and Y

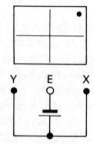
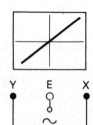

Fig. 6.13 Results of a.c. and d.c. inputs to an oscilloscope

the length of the line indicates the full swing of the voltage, e.g. +5 V to −5 V. Scales can be fitted to convert the deflections directly into voltages.

6.11 The timebase control

Although deflections of a stationary spot of light are sometimes used, the more common use is to arrange for the spot to traverse the

(*a*) d.c.

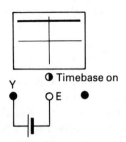

(*b*) a.c., frequency 100 Hz (100 cycles per second)

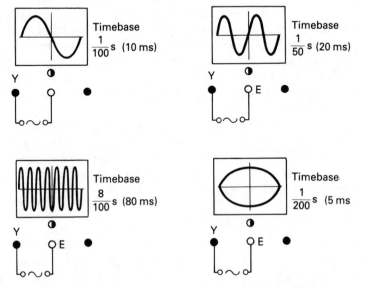

Fig. 6.14 Oscilloscope traces with timebase applied to X-plates

screen from left to right repeatedly, flying back between each sweep in a very short time and invisibly. The circuit which controls this type of movement is called a *timebase* and the time it takes the spot to sweep across the screen can be varied from, say, two seconds to $\frac{1}{2000}$ second. When the timebase is on, no other voltage can be applied to the X-plates.

The a.c. and d.c. displays with timebase on are shown in Figure 6.14. To get a steady trace using a.c., the repetition time of the timebase circuit must be adjusted to synchronise with the repetitions of the a.c., so that the spot traces out the same path on each sweep of the screen. Can the reader explain how the last trace of Figure 6.14 occurs?

6.12   The CRO as a voltmeter

Since the deflection of the spot is the effect of a voltage applied to the X- or Y-plates, its movement from the zero position can be used as a measure of the applied voltage. All that is needed is a knowledge of how far the spot moves for, say, five, ten, fifteen, twenty volts and a scale on the screen to convert the deflection into volts. Most CROs have this calibration done in manufacture, so that the Y-gain control (Figure 6.12) will have a range of settings marked in volts per cm with multiples indicated for each setting, e.g. $\times 1$, $\times 5$, $\times 10$, $\times 100$, etc. With a transparent scale marked in cm and mm, any position of the spot can be converted into volts with the appropriate factor being used from the position of the Y-gain control. If this spot moves 4.6 cm on the $\times 5$ setting, the voltage would be $4.6 \times 5 = 23$ V.

For a.c., similar measurements can be made from the height of the trace in Figure 6.14 or the length of the line in Figure 6.13. The value obtained gives the peak-to-peak voltage which could be converted into the more usually quoted root-mean-square (r.m.s.) value (Chapter 17 deals with this point in more detail).

6.13   Other uses of the CRO

The timebase control enables the CRO to be used for displaying changing voltages and for comparing different voltages visually, particularly alternating ones. Some of its applications are listed below:

comparison of input and output from amplifier or rectifier;
examination of a.c. voltage waveform;
display of musical waveforms using a microphone;
exploring voltages at different points in an electronic device;
display of physiological rhythms such as heart or brain activity.

The television tube is a special type of CRO in which the spot of light traverses the entire area of the screen in 625 interlaced lines 25 times each second. Signals from the TV aerial are made to affect the brightness of the beam and so produce a picture. Colour pictures require three separate electron beams which impinge on spots of different fluorescent material arranged over the inside of the tube, mixtures of three colours of light emitted allowing the effect of full colour to be produced.

6.14 Summary

A galvanometer is a current indicator.

Galvanometers can be converted to ammeters by adding shunts (resistors).

Galvanometers can be converted to voltmeters by adding multipliers (resistors).

Ammeter conversions need to have low resistances.

Voltmeter conversions need to have high resistances.

Ohmmeters can be made from galvanometers and cells.

Multimeters comprise a variety of current, voltage and resistance ranges, covering d.c. and a.c.

Cathode ray oscilloscopes offer measurements and displays of voltages for d.c. and a.c.

7

Electromagnetism

Elements of magnetism

7.1   Magnetic poles and fields

The simple observation that the effects of a piece of magnetised material appear to be concentrated around its ends led to the idea of magnetic *poles*. When it was also discovered that a suspended magnet invariably settled down in line with a north-south direction (or very near to it), the names North and South were given to the poles which pointed those ways. The original name was 'north (or south) seeking pole' (Figure 7.1(*a*)). It was soon realised that pieces of magnetic materials sometimes attracted and sometimes repelled one another. Again the effects seemed to originate between the poles that were nearest together, and the rule that 'like poles repel but unlike poles attract' became the way to describe this behaviour (Figure 7.1 (*b*)).

In order to explain the action of magnets on one another, even when they are not in contact, the idea has been developed of a *magnetic field* as the region around a magnet where its influence was felt. (The notion of a field is a common one in physics, being useful in gravitation, light, electricity and nuclear physics as well as in magnetism. It arises in any situation where effects are observed at a distance from the causes of them.) The magnet's region of influence is where the field acts, in theory everywhere around a magnet but getting weaker further away from it. Thus it is reasonable to think of a magnetic field as having strength (i.e. a size which could be measured) and also direction, since forces are involved which certainly are directional (or vector) quantities.

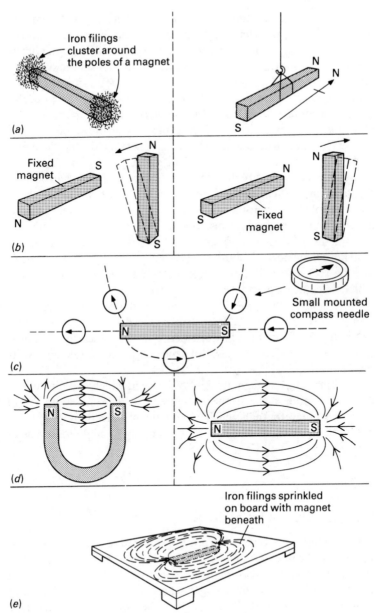

Fig. 7.1 Basic magnetic effects

What do magnetic fields look like? Imagine a small compass needle placed in the vicinity of a magnet. Wherever it is, its north pole will be attracted to the magnet's south pole and its south pole to the magnet's north, coming to rest under the action of these attractions as well as under the action of the repulsions between similar poles (Figure 7.1 (*c*) shows some positions). Moving the compass needle around from place to place enables a pattern to be seen, indicated by the direction in which the needle settles at each point. The pattern of lines depicts the magnetic field – the closeness of the lines gives a measure of the field's strength and the directions of the lines show the direction of the field. Because there will be only one field direction at every point, the field lines are never shown to cross or touch one another. Figure 7.1 (*d*) shows the field pattern around a straight magnet and also a horseshoe one. (Really, of course, the field is three-dimensional.) Note the direction of the field lines from north to south poles. Field patterns such as these can also be seen by using iron filings sprinkled over a board beneath which a magnet is placed (Figure 7.1 (*e*)).

Among the many uses of permanent magnets are: cupboard door catches, refrigerator door strips, screwdriver blades, focusing magnets for TV picture tubes, electric motors, bicycle dynamos, direction-finding compasses, loudspeakers, ammeters and volt-meters.

The Earth has a magnetic field of its own, which explains why a compass needle settles in the N–S direction. The shape of its field is very similar to that which would be produced by a straight magnet near to the Earth's centre but at a slight angle to its axis of rotation.

7.2 Combinations of fields

The relatively simple patterns around single magnets become much more complicated when more than one magnet is present and when the Earth's field is also involved. Figure 7.2 shows several such patterns. The important thing is that in some places the fields reinforce one another, whilst in others they weaken or even cancel one another out (e.g. at points marked with a cross).

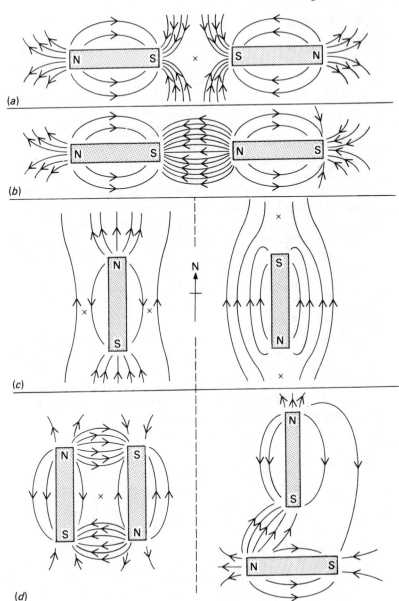

Fig. 7.2 Combinations of magnetic fields; (*a*) (*b*) (*c*) between two magnets; (*d*) between a magnet and the Earth

7.3   Magnetic materials

The magnetic rock, lodestone, is an ore of iron, but of the pure metals, only iron, cobalt and nickel can be magnetised. Alloys of these three with other metals such as aluminium, or steels containing carbon in varying amounts, make very powerful permanent magnets. (Stainless steel, however, is usually not magnetic because of the chromium content.) Other man-made materials can be formed into ceramic magnets or rubber magnets, also with great magnetic strength but often smaller physical strength. All this group of materials are known as *ferromagnetics* and show strong responses to magnetic fields. It may be surprising to learn that all other substances, even liquids, are in fact very slightly magnetic, but their properties are important only in very special circumstances.

A magnet has two main characteristics – the strength of its magnetisation and the degree to which it is retained. For use as a permanent magnet it is more important for a material to retain its magnetism, whereas for electromagnets (section 7.6) or transformers (Chapter 10) it is the ability to be switched on and off easily which matters more.

Ceramic magnets provide interesting playthings because of their lightness and high magnetisation. They are found in bar form, or rectangular or ring shapes, and can be magnetised in unexpected ways (Figure 7.3). It is even possible for a ring magnet to be fully magnetised but for no poles to be apparent at all – can you imagine how it might be magnetised?

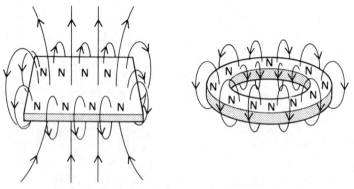

Fig. 7.3   Fields near ceramic magnets

Fields and currents

7.4 Magnetic fields from electric currents

The fact that an electric current produces a magnetic field can be shown by holding a compass needle over a wire in which there is a current. When the current is switched on, the needle is deflected from its normal N–S position. The shape of the field can be investigated using a compass or iron filings, as Figure 7.4 shows.

The field pattern is a series of concentric circles, the strength getting weaker further away from the current. Reversing the current causes the field direction to reverse. From a current of, say, 1 A the field is not very strong, but with larger currents quite strong fields can be generated.

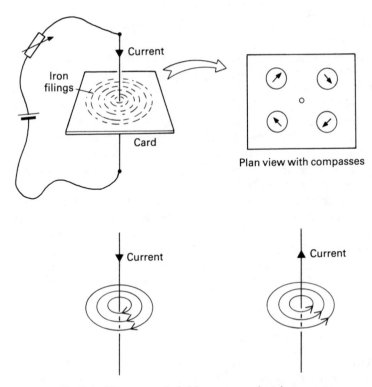

Fig. 7.4 The magnetic field near to an electric current

7.5 Current loop and solenoid

If a conductor carrying an electric current is formed into a loop, a more interesting field pattern is formed (Figure 7.5).

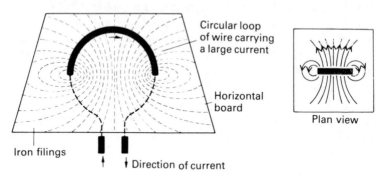

Circular loop of wire carrying a large current

Horizontal board

Plan view

Iron filings

Direction of current

Fig. 7.5 Field pattern through a loop

Note the field in the centre of the loop; the effect can be taken further by winding the wire carrying the current into the shape of a *helix* or *solenoid*. In this case the field patterns both inside and outside the solenoid are important (Figure 7.6).

Inside the solenoid the field pattern is a set of evenly spaced straight lines. Its strength depends on the size of the current and the closeness of the turns of wire. By winding a few hundred turns it is possible to produce powerful and uniform fields inside a solenoid which are easy to control by means of the current flowing.

The external field pattern looks remarkably like that of a permanent magnet (Figure 7.1 (*d*)). In fact the two would be indistinguishable if the fields were the only factor to be used for comparison. The solenoid even behaves as if it had north and south poles which repel and attract others just like magnets' poles. Left to swing freely (not easy to do because of the need to have wires connected to it) the solenoid points north–south.

7.6 Electromagnets

The internal magnetic field of a solenoid can be used to make an *electromagnet* if the coils of the solenoid are wrapped around an iron core. The solenoid's field magnetises the iron, which then adds its

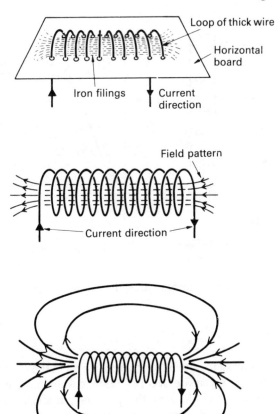

Fig. 7.6 Magnetic fields inside and outside a solenoid

much greater field to that of the electric current. The usefulness of this arrangement is that, provided the core is an iron one (i.e. almost pure iron, not a steel), its magnetism is not retained when the current is switched off. We have, in effect, a magnet that can be turned on and off.

The strength of the electromagnet depends on the current in the solenoid, the number of turns of wire and the volume of the iron core. Figure 7.7 shows a simple way of investigating these factors.

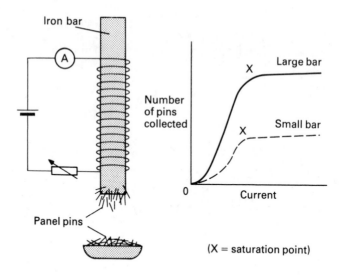

Fig. 7.7   Investigating the strength of an electromagnet

The important thing to note is that there is a limit to the electromagnet's strength for a given size of core – reached when the iron is fully magnetised or *saturated*. Beyond that point no purpose is served by trying to obtain greater effect. As far as the magnetising solenoid is concerned, it is the number of *ampere-turns* which matters, i.e. 3 A and 100 turns is just as effective as 2 A and 150 turns or 1 A and 300 turns.

7.7   Types of electromagnets

Several types of electromagnet design are shown in Figure 7.8, each appropriate for particular purposes. Care needs to be taken when using more than one coil of wire that the windings are in the correct sense to produce two different poles. (The drawings are not to scale.)

As with permanent magnets, there is a host of applications of electromagnets: bells and buzzers, separators of iron/steel from

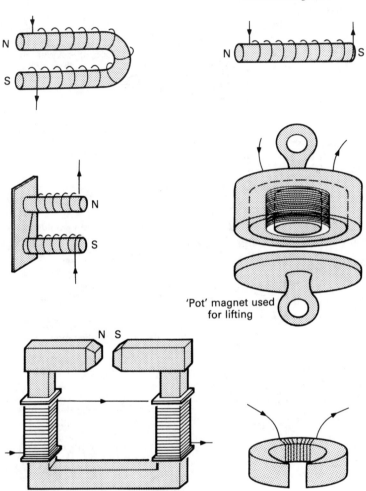

Fig. 7.8 Types of electromagnet

other materials, tape recorders, starter relays in motor cars, transformers, electricity generators, electric motors, telephone earpieces, inductions coils, atomic particle accelerators (e.g. cyclotron). Some of these uses are illustrated in Figure 7.9.

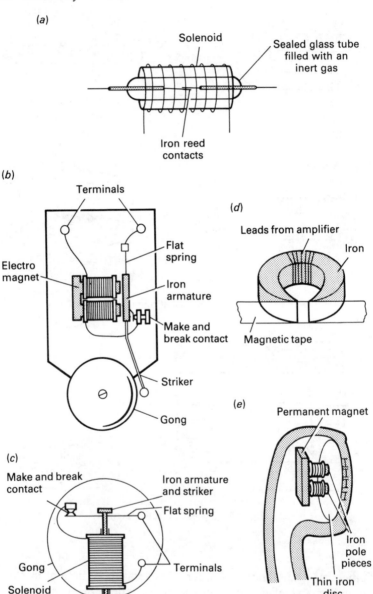

(a)   Solenoid   Sealed glass tube filled with an inert gas

Iron reed contacts

(b)   Terminals

(d)   Leads from amplifier   Iron

Flat spring

Electro magnet

Iron armature

Make and break contact

Striker

Gong

Magnetic tape

(e)   Permanent magnet

(c)   Make and break contact   Iron armature and striker

Flat spring

Gong

Solenoid   Terminals

Iron pole pieces

Thin iron disc

Fig. 7.9   Some applications of electromagnets

(a) A 'reed switch' where contacts which are part of one circuit are brought together by the magnetic field produced by current in another circuit. This is a type of relay where one current is used to switch on another one.

(b) The standard electric bell which depends on the attraction of a soft iron 'armature' to an electromagnet, which causes the circuit to be broken, the armature to spring back and the sequence of events to be repeated. The adjustment of the contact is critical to the smooth ringing of the bell.

(c) Another design of electric bell. The armature is drawn up into the . . .. Can you work out what happens to keep the bell ringing?

(d) The recording (or playback) head in a tape recorder produces a pattern of magnetisation in the particles of the tape under the influence of the electrical signals from the amplifier.

(e) The earpiece of a telephone contains a thin disc which vibrates according to the changing magnetic field produced by an electromagnet.

7.8 Summary

Similar magnetic poles repel one another, dissimilar ones attract.

Magnetic fields exist around magnetised materials or electric currents.

Lines of magnetic field do not touch or cross one another.

The Earth has a magnetic field of its own.

Different magnetic materials have different strengths and retention properties.

Magnetic fields around solenoids are similar to those around straight magnets.

The strength of an electromagnet depends on the size of the core and the number of ampere-turns.

8

Electrical Energy and Power

8.1 Energy and power

Energy is usually measured in the standard scientific unit, the joule, named after the British experimenter James Prescott Joule (d. 1889) who investigated the relationship between mechanical energy and heat, around 1850. In some countries and for some purposes in Britain an older unit, the calorie, is still preferred, or the larger Calorie (1000 calories). The connection between them is that 1 calorie = 4.18 joules, or 1 Calorie = 4.18 kJ.

A joule is the amount of energy transferred from one form to another when a force of one newton moves through a distance of one metre. Although defined in mechanical terms, the joule is applicable to all kinds of energy. To get a feel for the size of a joule, it is the energy transfer made when lifting a mass of 98 grams (about the size of a small chocolate bar) through 1 metre against gravity (Figure 8.1) or that when lifting a kilogram bag of sugar through 10.2 cm (just over four inches).

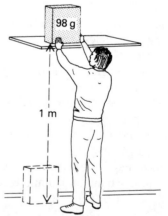

Fig. 8.1 One joule of energy is transferred by lifting 98 g through one metre

It is sensible to talk about the *energy transfer* or work done during a single event like lifting a weight through a certain distance. When dealing with a continuous action like the running of a car engine or a blast furnace in operation, however, it is more useful to consider the *rate* at which energy is being transferred or work done, i.e. the number of joules per second. Another name for a joule per second is a *watt* (named after James Watt, d. 1819) and it is the unit in which *power* is measured, power being the rate of working or the rate of transfer of energy. The abbreviation for watt is W.

		1 watt	$\equiv$	1 joule per second
1 kW	=	1000 W	$\equiv$	1000 Js^{-1}
1 MW	=	1 000 000 W	$\equiv$	1 000 000 Js^{-1}
1 mW	=	$\dfrac{1}{1000}$ W	$\equiv$	$\dfrac{1}{1000}$ Js^{-1}

These statements are not equations in which left and right hand sides have equal values; they are indentities in which left and right hand sides are the same things given different names.

Occasionally the old unit of power, *horsepower* (h.p.), is still met, at least in conversation. It is equal to approximately 750 W. A 70 kg (11 stone) man who runs up a flight of 12 steps, each 23 cm (9 in) high, in 5 seconds would be working at a rate of about 375 W or ½ h.p.

8.2 Amps, volts, ohms and watts

When one coulomb moves through a p.d. of one volt, one joule of energy is transferred (section 3.3).

So, if 1 C moves each second through 1 V, 1 J is transferred each second.

But, 1 C moving each second is 1 A (section 1.4).

So, if 1 A passes through a p.d. of 1 V, 1 J is transferred each second.

But, 1 J every second is 1 W (section 8.1).

So, if 1 A passes through a p.d. of 1 V, the rate of energy transfer is 1 W.

So, if 2 A passes through a p.d. of 3 V, the power transfer is $2 \times 3 = 6$ W.

So, if a current of I passes through a p.d. of V, the power transfer is IV.

Rate of transfer of energy
or *electrical power* = Current × Potential
(in watts)   (in amps) difference
(in volts)

In symbols: $P = IV$

It should be stressed that this relationship between power, current and potential difference is not a clever bit of obscure physics: it follows directly from the way the volt is defined as a joule per coulomb.

Using the relationship $V = IR$ (section 4.8), two other equations can be formed by substitution for I or V:

$$P = I^2R \qquad P = \frac{V^2}{R}$$

8.3   Calculations on power

Any of the three equations for P can be used according to the information given and the result to be calculated.

Example 1   What is the rate of energy transfer, i.e. power, when a current of 5 A passes through a p.d. of 12 V?
         Using $P = IV$, power = 12 W = **60 W**

Example 2   A current of 4 A flows through a 20 Ω resistor. What is the power?
         Using $P = I^2R$, power = $4^2 \times 20$ W = **320 W**

Example 3   A 10 Ω resistor has a p.d. of 100 V across it. What is the power?
Using $P = \frac{V^2}{R}$, power = $\frac{100^2}{10}$ W = **1000 W**

Note the language used in these examples – power is the rate of energy *transfer* rather than energy dissipated or lost or used. To pass a current through a resistor the supply has to convert energy at the rate worked out, and the energy is transferred from electrical into thermal because the resistor will heat up in the process. Certainly as far as the supply is concerned the energy is 'lost' to it, but the energy does not vanish completely. The supply has to work at a certain rate in order to effect the energy changes each second.

Example 4 A power pack has an internal resistance of 12 Ω and sends a current of 10 A through a resistor of 20 Ω. How much power does the pack supply and what proportion of it is involved in heating up the resistor?

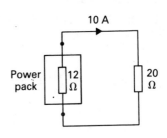

Total resistance = (12 + 20) Ω
= 32 Ω
Using $P = I^2R$, for the whole circuit, total power = $10^2 \times$ 32 W = **3200 W**
Using $P = I^2R$ for the external resistor, power involved in the resistor = $10^2 \times 20$ W = 2000 W
Porportion of power involved in heating the resistor $= \dfrac{2000}{3200} = \dfrac{\mathbf{5}}{\mathbf{8}}$

Example 5 In the circuit shown, at what rate is the power supply delivering energy to the external resistors?

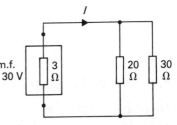

Equivalent external resistance R is calculated from:

$$\frac{1}{R} = \frac{1}{20} + \frac{1}{30} = \frac{3 + 2}{60} = \frac{5}{60}$$

$$R = \frac{60}{5} \ \Omega = \mathbf{12\ \Omega}$$

For this equivalent 12 Ω resistor we need to know either the current flowing through it or the p.d. across it; the current is more readily found:

$$\text{Using } E = Ir + IR$$
$$30 = I(3 + 12)$$
$$30 = I \times 15$$
$$I = 2\ \text{A}$$
$$\text{Using } P = I^2R \text{ for the external resistors,}$$
$$\text{power} = 2^2 \times 12\ \text{W} = \mathbf{48\ W}$$

Beware of trying to use $P = V^2/R$ here, because not all the e.m.f. of 30 V is applied to the external resistor; some of it is needed to send current through the internal resistance, 6 V in this case ($V = Ir$). This example, with different quantities involved, can be used to show that the power delivered to the external resistor is greatest when internal and external resistances are equal – an important result which applies to any source of electricity.

Example 6   An electric lamp is sold as 240 V, 100 W. What is its resistance when in use and how much current does it take?

$$\text{Using } P = \frac{V^2}{R}, \quad 100 = \frac{240^2}{R}$$

$$R = \frac{240^2}{100} \ \Omega = \textbf{576 } \boldsymbol{\Omega}$$

$$\text{Using } V = IR, \quad 240 = I \times 576$$

$$I = \frac{240}{576} \ \text{A} \ = \textbf{0.42 A}$$

$$\text{Or using } P = IV, \quad 100 = I \times 240$$

$$I = \frac{100}{240} \ \text{A} \ = \textbf{0.42 A}$$

8.4   Paying for electricity

Thinking of what electricity is used for in the home or in commerce and industry – heating, lighting, driving machinery, sound equipment, electromagnets, chemical plants, transport – it is clearly energy that is being provided.

One joule of energy would not be enough to heat a thimbleful of water to any noticeable extent, so a much larger unit of measurement is needed – kilojoules or megajoules, even. A 100 W lamp running for one minute would transfer $100 \times 60 = 6000$ J of energy. Another way of expressing this would be to call it 100 watt-minutes, but even that is a relatively small quantity of energy. In practice, a common unit is the kilowatt-hour (kWh).

```
 1 kWh means  1000 Watts for 1 hour
          or  1000 joules per second for 1 hour
          or  1000 × 60 × 60 joules
 1 kWh  =  3 600 000 J
        =  3.6 MJ
```

The kWh or MJ are both conveniently sized quantities of energy for domestic use and the ordinary electricity meter records one or other of them as the 'units' of electrical energy for which charges are made.

8.5   Costing particular items of equipment

All electrical appliances carry labels on which the manufacturer shows the power ratings and the voltages for which they are designed. In Great Britain, 240 V is the standard domestic voltage supplied through the grid (section 13.3), though industrial premises might also use a 415 V supply. 110 V is common in Europe and the United States.

Some consumer goods and typical power ratings are listed below together with the costs of running them for an hour at N pence per kWh.

Appliance	Power rating	Cost in pence for one hour
Shaver	8 W	0.008 N
Lamp	100 W	0.1 N
Electric blanket	120 W	0.12 N
Music centre	140 W	0.14 N
Television set	190 W	0.19 N
Hedge trimmer	400 W	0.4 N
Coffee maker	725 W	0.725 N
Hair dryer	1.2 kW	1.2 N
Electric fire (two-bar)	2.0 kW	2.0 N
Kettle	2.4 kW	2.4 N
Immersion heater	3.0 kW	3.0 N
Cooker and four rings	11.0 kW	11.0 N

The most expensive appliances to run are those which incorporate heating elements. Immersion heaters, in particular, are easy to leave switched on for long periods of time – an automatic time switch is certainly a good investment for this item. The figures for the electric cooker are misleading because they do not allow for thermostatic controls which are fitted to most ovens and heating rings, switching them on and off to maintain particular temperatures or rates of cooking. Similarly, the rate of energy transfer in a television set or music centre is not a steady amount.

8.6   Summary

Power is the rate of transfer of energy and is measured in watts.
A watt is a joule per second or a volt-ampere.

$$P = IV \quad I = \frac{P}{V} \quad V = \frac{P}{I} \quad P = I^2R \quad P = \frac{V^2}{R}$$

Units of electrical energy are costed in kilowatt-hours or megajoules.
A kilowatt-hour (kWh) is 3.6 MJ.

9

Electric Motors

9.1 Movement out of electromagnetism

In Chapter 7 various magnetic field or flux patterns were described
and illustrated from both permanent magnets and electric currents.
The interaction of fields from different sources can produce forces,
and therefore movement, if the right arrangements are designed to
take advantage of them.

The basic motor effect arises from the combined effects of the
field of a magnet with that of a current.

Figure 9.1 shows an experimental arrangement by which a current can be made to pass in the field between the poles of a strong
magnet. The bar carrying the current can roll sideways along a pair
of brass rails through which the current is conducted. When the
circuit is switched on, the bar rolls sideways due to a force generated
by the magnetic fields of the magnet and the current. Figure 9.2
shows the separate field patterns and their combined one.

The force arises from the very unsymmetrical pattern and both
the magnet and the bar carrying the current experience forces, in
opposite directions, which try to even out the field. The direction
of the force depends on the directions of the magnetic field of
the magnet and the current. Figure 9.3 summarises the 3-dimen-
sional relationship between these directions and shows the force
experienced by the conductor carrying the current.

In order to remember which directions are which, it might be
useful to use Fleming's left-hand rule (Figure 9.4).

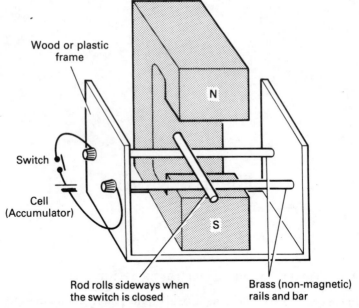

Wood or plastic frame

Switch

Cell (Accumulator)

Rod rolls sideways when the switch is closed

Brass (non-magnetic) rails and bar

Fig. 9.1 An arrangement to show the interaction of the magnetic fields due to a magnet and a current

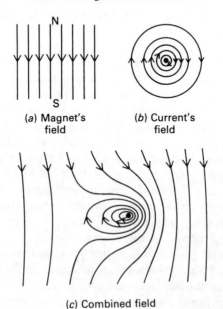

(a) Magnet's field

(b) Current's field

(c) Combined field

Fig. 9.2 Combination of fields from magnet and current

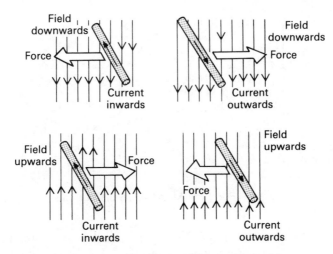

Fig. 9.3 Relative directions of field, current and force

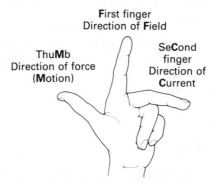

Fig. 9.4 Fleming's left-hand rule

9.2 Forces between currents – the ampere

The left-hand-rule way of working out what happens can be used to decide how the forces act when two currents in parallel straight wires are close to each other (Figure 9.5). The curious result is that currents in the same direction attract each other, but currents in opposite directions repel each other.

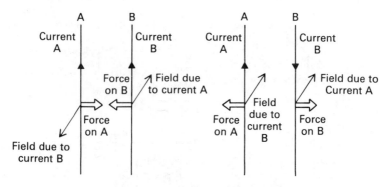

Fig. 9.5 Forces between two parallel currents

An important definition of the *ampere* makes use of the force between two current-carrying conductors:

> An ampere is that current which when flowing through two infinitely long thin parallel conductors spaced 1 m apart in a vacuum produces a force between them of 2 × 10^7 N per metre length of the conductors.

Although only a theoretical definition, it is possible to measure currents by an elaborate 'current balance' which is based on the definition, and is the internationally agreed way of fixing the meaning of an ampere.

9.3 The Hall effect

In a case of a magnet's and a current's field interacting where neither the magnet nor the conductor could move, there would still be a force on the electrons within the conductor, which would shift more of them to one side as they move along. The result of this is a small p.d. developed *across* the direction of the current – a phenomenon known as the *Hall effect*. Figure 9.6 illustrates the relationship between the factors involved. Semi-conducting materials like germanium show a large Hall effect and they can be used to measure the strength of magnetic fields from measurements of the p.d. developed and the current.

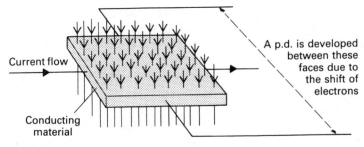

Fig. 9.6 The Hall effect

9.4 A simple d.c. motor

Figure 9.7 illustrates a way of making a simple d.c. motor which has many aspects of design in common with large commercial motors. An *armature* (wooden block) carries a *coil* of insulated wire with several *turns* and can spin freely in a *magnetic field*. A method of leading the current into and out of the coil through contacts is provided which is not inconvenienced by the rotation of the armature.

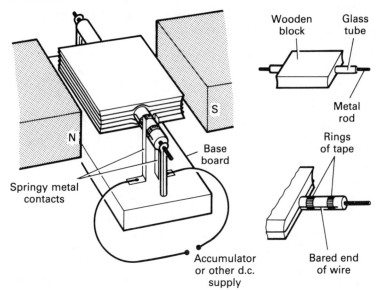

Fig. 9.7 Making a simple d.c. motor

The motor works because opposite sides of the coil experience forces in opposite directions, causing the armature to spin. Figure 9.8 shows the forces acting (*a*) when the coil is horizontal and (*b*) at an angle between the poles of the magnet.

If the coil were to spin further with the forces, as shown, it would quickly come to rest in a vertical position, with the forces on AB and CD keeping it in that place. What is needed to maintain the rotation is for the current direction around the coil to be *reversed* as it passes through the vertical, as shown in Figure 9.8(*c*), which gives forces able to spin the coil further. If the current direction can be reversed every half-revolution as the coil passes the vertical (i.e. parallel to the faces of the pole pieces of the magnet), the forces also reverse and the coil is kept rotating.

The device which causes the reversals of current is called a *commutator*. Figure 9.9 shows details of a simple commutator, in which carbon brushes make contact with two insulated half-cylinders, to which the coil is connected. This is an improvement on the method used in Figure 9.7 because electrical contact can be maintained throughout the rotation of the coil rather than intermittently, only as its wires touch the metal contacts. (In Figure 9.9 the carbon brushes would be fixed to the housing of the motor and the copper ring would rotate between them with the coil.)

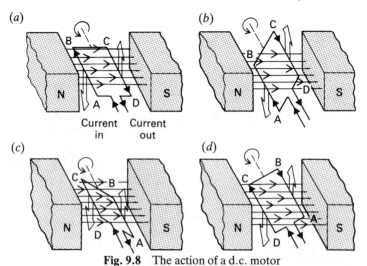

Fig. 9.8 The action of a d.c. motor

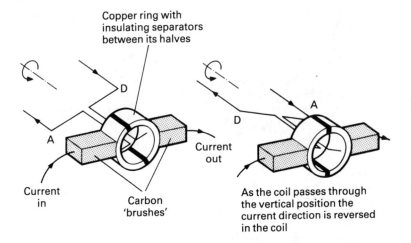

Fig. 9.9 Detail of the commutator

9.5 Some design features of practical d.c. motors

Real motors have a more elaborate and robust construction than the model described above. Properly engineered bearings, for instance, are needed with machined parts and protective casing, but increased efficiency is also obtained by having several coils of wire spaced out around a cylindrical armature and connected in series. A multiway commutator with carbon brushes provides the necessary connections and current reversals. The armature itself is made of laminated soft iron and the pole pieces of the magnet are curved to fit around it. Small motors used for toys have permanent magnets, but larger ones use electromagnets which need separate windings to energise them. Figure 9.10 shows the ways these features are arranged.

With the use of electromagnets in motors there are three ways in which they can be connected to use the same power supply as the armature (Figure 9.11). Each type of winding gives the motor particular properties and therefore particular applications.

A property of all electric motors, which will be considered in section 10.13, is that they take a very large current when first switched on but less as they pick up speed. It is as if the motor has a

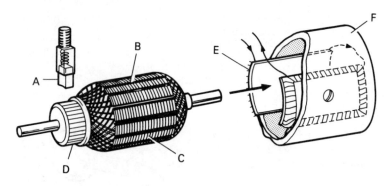

A   Spring-loaded replaceable carbon brushes.
B   Many coils, in series, connected to appropriate segments of the commutator.
C   Laminated soft iron armature, carries the coils in grooves.
D   Multi-way commutator, one pair of segments for each coil.
E   Heavy-duty coils for driving the electromagnet.
F   The electromagnet forms the outer case of the motor. Concave pole pieces ensure maximum torque on the armature.

Fig. 9.10   Construction of a practical d.c. motor

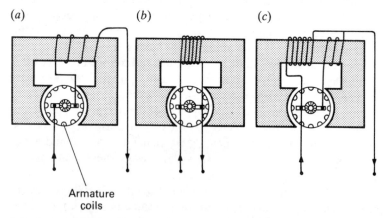

(*a*)   Series wound: field coil in series with armature coils.
(*b*)   Shunt wound: field coils in parallel with armature coils.
(*c*)   Compound wound: two field coils, one in series and one in parallel, with armature coils.

Fig. 9.11   Three types of windings for d.c. motors

speed-dependent resistance. The reasons for this are explained in Chapter 10, but note here that it is important to include a variable resistor (which can be reduced progressively) in a circuit containing a motor, to prevent undue heating of the windings (Figure 10.18).

9.6 a.c. motors

The most complicated part of a d.c. motor is the commutator, which reverses the current direction in the armature coils every half-revolution, as we have seen. If alternating current is used instead, such an arrangement becomes unnecessary because the current reversals take place automatically. The problem of maintaining contact remains, though. Figure 9.12 shows an adaptation of the simple motor design in which the ends of the coil are supplied from the same contacts all the time, instead of alternately from each as in the d.c. case of Figure 9.7.

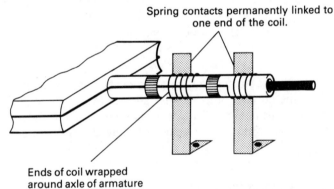

Spring contacts permanently linked to one end of the coil.

Ends of coil wrapped around axle of armature

Fig. 9.12 Details of contacts for simple a.c. motor

Since the rotation of an armature depends on the current changing direction at the right moment, there will be a certain speed at which an a.c. motor functions in accordance with the frequency of the alternating current. Such a *synchronous* motor has an obvious use in electric clocks designed to work off the 50 Hz mains supply. Figure 9.13 shows details of the construction of this type of motor. The inner and outer set of poles have their directions of magnetisation reversed each time the a.c. reverses. If the rotor spins at just the right speed, its permanent poles are continually attracted to the next

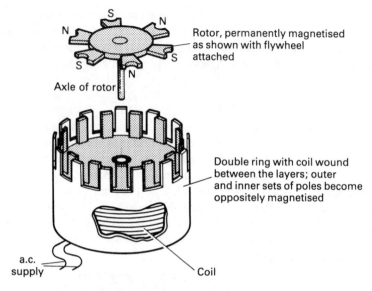

S

N   N

N   Rotor, permanently magnetised
                                          as shown with flywheel
S   attached
                          S

Axle of rotor   N

                                          Double ring with coil wound
                                          between the layers; outer
                                          and inner sets of poles become
                                          oppositely magnetised

a.c.
supply   Coil

Fig. 9.13   An a.c. synchronous motor

pair of fixed poles and a steady rotation can be maintained which is related to the number of fixed poles and the frequency of the a.c. supply. This type of motor usually needs a start by giving the rotor a gentle spin. A.c. motors for heavy duty are designed differently (section 10.11).

9.7   Galvanometer design

Another application of the motor effect is to be found in the construction of the *moving-coil galvanometer*, which can be adapted to serve as ammeter, voltmeter and ohmmeter (Chapter 6). Like a simple d.c. motor, there is a coil of wire and a magnetic field and therefore rotation occurs, but in this case springs are employed to prevent the coil turning through more than about 90°. Figure 9.14 shows details of the arrangement.

Note that the rectangular coil of wire on its aluminium former moves in the space between the magnet pole fields and the iron cylinder. The cylinder is fixed to the magnet by a non-magnetic

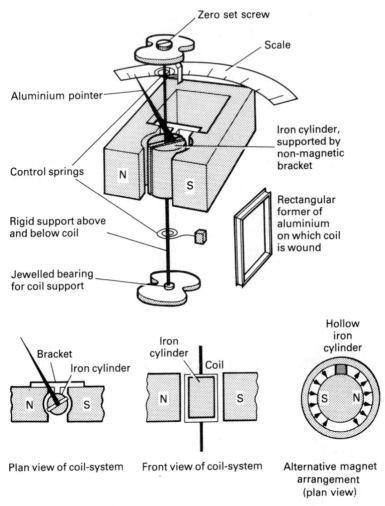

Fig. 9.14 Details of a moving-coil galvanometer

bracket. Current is led into and out of the coil through the control springs.

Note the use of an aluminium former on which the coil or wire is wound (see section 10.11), and the narrow gap between pole pieces and core in which the coil can rotate. A stiff wire suspension with jewelled bearings in the case of the instruments helps to make a robust design. The control springs above and below the coil also serve as contacts for the current. The permanent magnet which supplies the field can be placed inside the coil rather than outside it, as illustrated, which allows a larger angle of swing for the coil and also reduces the weight of the instrument by using a smaller magnet.

9.8   Sensitivity of a galvanometer

The angle of deflection of the coil in a galvanometer made as described above depends on:

the strength of the magnet B;
the size of the current I;
the number of turns of wire on the coil n;
the area of the coil A;
the stiffness of the control springs c.

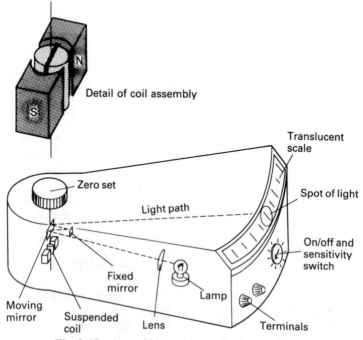

Fig. 9.15   A sensitive moving-coil galvanometer

To give a large deflection would need the first four in the list to be large and the fifth to be small. Factors such as size and weight will in practice limit how far B, n and A can be increased. Often n might be several hundreds of turns of fine wire but A only one or two cm^2. B depends on the alloy used for the, magnet and I is the current to be measured. It is the stiffness of the spring which offers most scope to determine the sensitivity of the galvanometer.

For very sensitive instruments a different but more delicate suspension system is used, which itself provides the control in place of the springs. Instead of a pointer, a beam of light and a system of mirrors can be arranged which produce an observed deflection of several cm for a current as small as one microamp. Figure 9.15 shows such a galvanometer. Care needs to be taken when moving these instruments not to damage the suspension. A clamp is often fitted to secure the coil in transit. Levelling screws and a spirit level may be needed to ensure proper alignment.

9.9 Moving-coil loudspeaker

Many loudspeakers connected to TV and radio receivers, tape recorders or public address systems are moving-coil devices. They are fed with alternating currents from amplifiers and the coils vibrate rather than turn or twist. Figure 9.16 shows how the coil and the magnetic field are arranged to produce a to-and-fro motion of

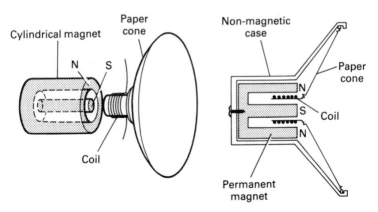

Fig. 9.16 The moving-coil loudspeaker

the coil. This is communicated to a paper cone which vibrates the air in front of it. It is interesting to consider how the magnet is magnetised to produce the poles as shown.

9.10   Summary

Combinations of electric currents and magnetic fields produce forces and movement.

The force acts at right angles to both field and current directions.

The ampere is defined in terms of the forces between currents.

A d.c. motor requires a commutator and an armature.

Laminated armature cores are used to reduce eddy currents.

A simple a.c. motor requires slip rings, and can run synchronously.

Moving-coil galvanometers are modified d.c. motors.

Moving-coil loudspeakers are a.c. devices for converting electrical signals into sounds.

10

Electromagnetic Induction

10.1 The 'opposite' of the motor effect

In Chapter 9 it was seen that movement could be produced out of a magnetic field and an electric current. A sort of opposite effect, in which electrical energy can be *generated* from movement and a magnetic field, is the subject of this chapter. (Many things in science seem to have 'opposites' in which an inverse effect can be produced: electrolysis and fuel cells; microphones and loudspeakers; piezo-electricity and quartz oscillator; the expansion of a wire when heated and its cooling when stretched.) In this case it is the motor effect and the dynamo or generator effect which form 'opposites', and some similarities will be noticed with much of what was described in the last chapter.

10.2 Simple experiments – Faraday's Law

The essential features of *electromagnetic induction*, as this process is called, can be appreciated from a series of experiments using quite basic equipment (Figure 10.1). The common features of these experiments are: (*i*) a magnetic field from a magnet or electromagnet; (*ii*) a separate wire or coil; and (*iii*) relative movement between them. In diagrams (*a*) and (*b*) the wire is shown being moved up and down, in (*c*) and (*d*) it is the magnets which are moved, and in (*e*) and (*f*) the coils. In all cases the galvanometer is deflected to and fro in step with the movement.

What exactly is induced by the movement? The galvanometer's deflection shows the existence of a current, but currents do not arise

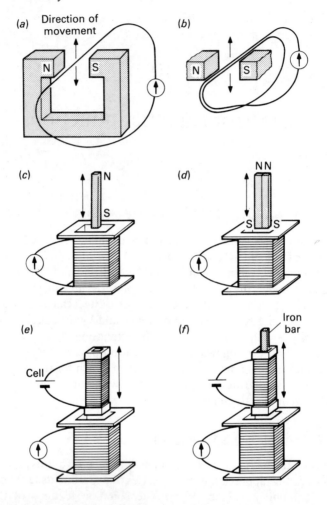

(a) A wire moved up and down causes a deflection of the galvanometer; sideways movement causes no deflection.
(b) More turns of wire give a larger deflection.
(c) A magnet moved into and out of a solenoid causes deflection.
(d) A stronger magnet and more turns each gives a larger deflection.
(e) A primary solenoid with current causes deflection when moved into and out of a secondary solenoid.
(f)   An iron rod greatly increases the effect.

Fig. 10.1   Simple induction experiments

by themselves – there has to be an electromotive force which makes the charges flow, so it is more accurate to say that an e.m.f. is induced. Whether any current results from the e.m.f. will depend on there being a circuit in which the charges can flow, and the current size will be governed by the resistance in the circuit. Investigations through these experiments were conducted by Michael Faraday who summarised his findings in the form of a law:

> When the magnetic flux through a circuit is changed, an e.m.f. is induced in the circuit which is proportional to the rate of change of the magnetic flux.

(Magnetic flux is the proper name for what we have called magnetic field. It is a three-dimensional quantity indicating field strength and direction in a certain region or volume of space.)

10.3 Other electromagnetic induction experiments

The main points which Faraday recognised are that there has to be a *change* in the magnetic field and that the e.m.f. depends on the *rate of change* of the field. Figure 10.2 shows some more elaborate arrangements where magnetic fields are changed. In each case the galvanometer is deflected only when the field is actually changing, and anything which produces a change will cause an e.m.f. to be induced. Changes can be made by switching on and off, for instance, so physical movement is not necessary. Note the effect of iron linking the two coils, which greatly increases the size of the e.m.f.

10.4 The induction coil

Figure 10.2(*c*) illustrates the basis of a device called an induction coil, in which a high voltage can be developed in a secondary circuit when the current in a primary circuit is switched off. Figure 10.3 shows how the effect can be used to energise an electric fence for cattle.

The primary circuit consists of a battery, a switch and a coil of wire wound on an iron rod or bar. (The capacitor is present to reduce sparking across the switch – section 10.13.) The primary coil is made of quite thick wire, so that when the circuit is switched on there is a sizeable current. The secondary circuit's coil is wound

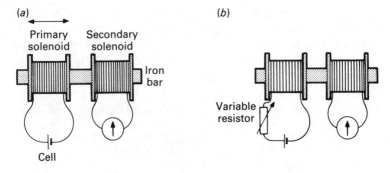

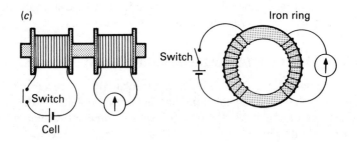

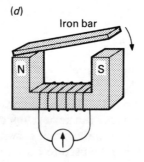

Changing fields can be produced by:
(*a*) moving the primary solenoid to and fro;
(*b*) altering the setting of the variable resistor;
(*c*) switching the primary circuit on and off;
(*d*) joining the magnet's pole with an iron bar.

Fig. 10.2 More induction experiments

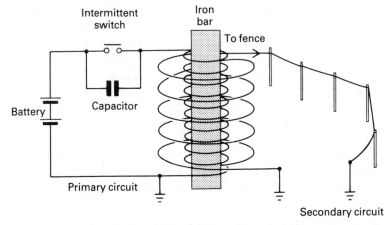

Fig. 10.3 Electric fence supplied from an induction coil

directly on to the primary (but using insulated wire, of course) and has many more turns of much thinner wire. The e.m.f. induced in this secondary coil is applied to the wire of the fence, which has specially insulated supports at the appropriate height for a cow's nose. Switching on the primary circuit gives a small induced e.m.f. in the secondary, but switching off has a much bigger effect due to the rapid magnetic change from full magnetisation of the iron bar to virtually zero. Several thousand volts can be developed, but for just a short time. A mechanism which repeatedly switches the primary current on and off will produce an intermittent high voltage to be applied to the fence, say once every three seconds, when the circuit is broken. An animal in contact with the wire of the fence will experience a short, mild, electric shock. Electric fences used to use small motors to operate the switch mechanically at the desired rate, but the switching can be done much more effectively by electric circuits. From the farmer's point of view electric fences have the great advantage of being easily moved from place to place.

A similar but more elaborate version of the induction coil is used to provide the ignition spark in motor car engines. A much higher voltage than the 12 V battery can provide is used for the sparking plugs, and it is obtained by inducing an e.m.f. in the same way as described above. A complication in this case is the need to synchronise the high voltages applied to, usually, four cylinders with the movements of the pistons, so that the petrol-air mixture explodes at just the right moment in each. Electronic systems can achieve this much more reliably than the previous contact-breakers.

Note that the purpose of the primary circuit in an induction coil is to provide a high degree of magnetisation of the iron core by using a large current. The wires for this circuit must be fairly thick to conduct the current. The secondary circuit on the other hand has an e.m.f. induced in each turn of wire, so the more turns it has the larger will be the effect. For many hundreds of turns the wire will need to be thin to avoid too much bulk, but the current is proportionately lower as well so thin wires are all that is necessary. Section 10.9 returns to this point in the context of the transformer.

10.5   The direction of the induced e.m.f. – Lenz's Law

As in the case of the electric motor, where there was a relationship between the directions of the field, the current and the induced force (Figure 9.1), so there is a corresponding one between the directions of the field, the movement and the induced e.m.f. By using an arrangement such as that of Figure 10.4 and relating the N–S field direction, the direction in which the wire is moved and the direction in which the galvanometer pointer is deflected, one can work out how they are connected.

Like the motor relationship, the three directions are mutually at right angles and can be remembered by using Fleming's *right*-hand rule (Figure 10.5).

You may be curious about the similar but different connection between the two electromagnetic effects – the motor and the induced e.m.f. Common sense, perhaps, might wish them to be the

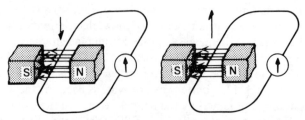

Fig. 10.4 Establishing the connection between field, movement and induced e.m.f.

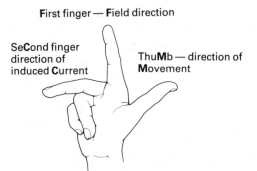

First finger — **F**ield direction

Se**C**ond finger direction of induced **C**urrent

Thu**M**b — direction of **M**ovement

Fig. 10.5 Fleming's right-hand rule

same, but there is a very important reason why they *must* be different. Consider again the basic motor effect shown in Figure 10.6.

The field is downwards from N to S, the current passes from A to B and the force acts to move the rod AB to the left. As soon as the rod moves, we have the basic induction effect of Figure 10.6(*b*) with the field again downwards, the movements of AB towards the left – but this means that the induced e.m.f. will be from B to A not A to B. Such an effect will reduce the current in AB and therefore slow down the motion of AB. If it were the other way round, with the induced e.m.f. from A to B, the effect would be to increase the current, making the rod move faster, giving a greater induced e.m.f., increasing the current further, moving the rod even faster,

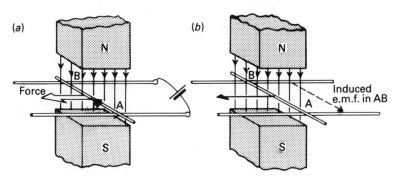

(a)

Force

B

A

N

S

(b)

B

A

N

S

Induced e.m.f. in AB

Fig. 10.6 The link between motor and dynamo effects

etc. If this happened we would be getting increasing energy out of the device merely for the price of assembling it and switching it on! Such energy-for-nothing machines do not exist and the reason in this case is that the induced effect has the result of slowing down the movement, not of enhancing it. This general result is expressed in Lenz's Law:

> The direction of the induced current is always such as to oppose the change which produces it.

Notice that Lenz's Law refers to the induced current, not the e.m.f. If the coil circuit had been left open, there would still have been an e.m.f. induced but no current could have flowed and therefore no opposing force could have been generated (see also section 10.11).

10.6   Transformers

Probably the most useful application of electromagnetic induction apart from the generation of electrical energy is the *transformer*. The principle of it is shown in Figure 10.7, which is similar to one of the arrangements shown in Figure 10.2.

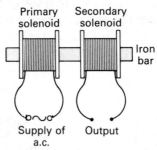

Fig. 10.7   The principle of the transformer

An induced e.m.f. arises only when there is a change in the magnetic flux through a coil. The change can be brought about by movement or by switching on and off, but if the field is produced from an alternating current there will be continuous changes both in its size and its direction. This results in a continuously changing induced e.m.f. The transformer is therefore an a.c. device in which an alternating current input gives an alternating e.m.f. output.

The closer the two coils can be put to each other, the more efficiently will the field of the *primary* link with the *secondary*, i.e. very little flux is 'wasted'. A closed system of iron is the best to make maximum use of the magnetic field. Figure 10.8 shows some practical designs.

The iron 'core' on which both transformer coils are wound is

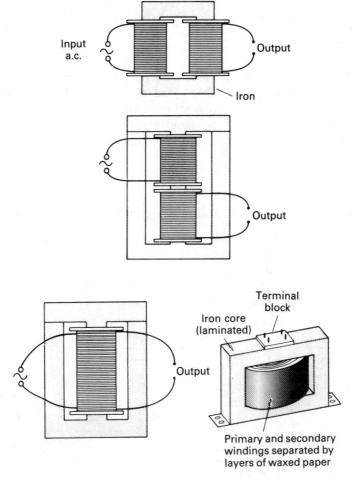

Fig. 10.8 Improved designs for transformers

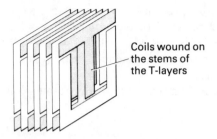

Coils wound on the stems of the T-layers

Fig. 10.9   Construction of a transformer core from U-shaped and T-shaped (shaded) layers of iron

made not of solid iron but of *laminations*, i.e. layers of iron insulated electrically from one another by coats of varnish before being riveted together. Figure 10.9 shows a typical construction. The reason for this design is to avoid loss of energy through eddy currents circulating in the iron core (section 10.10).

10.7   Step-up and step-down transformers

The wide use of transformers in electrical and electronic devices is due to their ability to change the level of an alternating voltage either up or down. The number of turns of wire on the primary and secondary coils determines the way the voltage is affected. For a 100 per cent efficient transformer:

$$\frac{\text{a.c. voltage input}}{\text{a.c. voltage output}} = \frac{\text{number of turns on primary coil}}{\text{number of turns on secondary coil}}$$

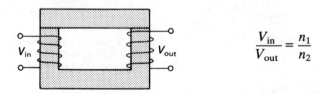

$$\frac{V_{\text{in}}}{V_{\text{out}}} = \frac{n_1}{n_2}$$

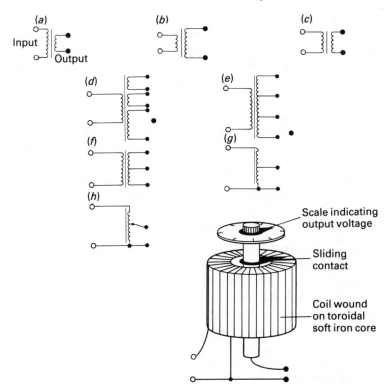

(a)   Step-down transformer used in battery charger, for example.
(b)   Step-up transformer.
(c)   1:1 transformer used for isolating one section of an a.c. circuit from another.
(d)   Transformer with three separate secondary windings: used for providing appropriate voltages for different components in TV or radio receivers.
(e)   A tapped transformer: provides steps of voltages, as used in laboratory power supply units.
(f)   A centre-tapped transformer: gives two equal voltages, used in rectifiers (Chapter 12).
(g)   An auto-transformer, uses only one winding: commonly found in '110 V/240 V' units such as electric shaver points.
(h)   A variable auto-transformer: used for providing any output voltage up to a limit of the maximum input voltage.

Fig. 10.10   Types of transformer and their applications

A step-up transformer increases the voltage, so has more secondary turns than primary, whilst a step-down one reduces the voltage and has more primary turns than secondary. A one-to-one turns ratio $(n_1 : n_2)$ will not alter the voltage but is useful for isolating one section of a device from another. More complex transformers with two or more secondary windings or with a tapped secondary winding are frequently used. Figure 10.10 shows circuit symbols for transformer types and the caption lists some of their applications.

Remember that transformers work only with alternating voltages. The induction coil (section 10.4) is a d.c. device, but it needs a mechanical switch to produce the necessary changing field and it gives a very 'spiky' intermittent output voltage.

10.8   The importance of transformers

Their ability to change a.c. voltages, up or down, puts transformers among the most widely used electromagnetic devices. In Chapter 11 their role in the distribution of electrical energy will be seen to be crucial, but they are equally valuable in domestic and electronic equipment. The very high efficiency of transformers also marks them out for many applications, few other machines being able to match their ninety per cent-plus efficiency. Many mechanical systems, such as cranes and winches, and most engines, such as locomotives and motor cars, cannot manage even forty per cent efficiency.

With such a wide variety of uses there is a wide variety of sizes of transformers, ranging from just over 1 cm across in electronic equipment to 5 m in power stations.

10.9   Transformers and power

A device made of copper wire and iron cannot develop energy of its own, so the power output of a transformer will never be greater than the power input. In the ideal case the input and output power would be equal, i.e.:

$$I_{in} \times V_{in} = I_{out} \times V_{out}$$

but this assumes one hundred per cent efficiency and no energy 'losses' in the process. In practice there are unavoidable energy changes which reduce the efficiency from one hundred per cent:

(a) copper losses – heat energy is produced by the current in both primary and secondary coils – I^2R for each coil (section 8.2).

(b) iron losses – heating caused by eddy currents (section 10.10) in the core and by the continuous changes in magnetisation of the iron itself.

(c) flux losses – the 'escape' of magnetic flux between primary and secondary coils.

The output power is therefore somewhat less than the input, but a carefully designed transformer can achieve up to ninety-four per cent efficiency.

A step-up transformer will increase the voltage, but if the product IV is to stay the same the available current must be decreased. For this reason the wire used in the secondary winding of a step-up transformer need not be as thick as that on the primary side.

10.10 Eddy currents

Electromagnetic induction occurs when a changing magnetic flux generated by a magnet or an electric current induces an e.m.f. in another circuit or conductor.

The induced e.m.f. causes a current and the effect is used in generators, dynamos and transformers. However, an e.m.f. will be induced in any conductor within the changing flux, not only the turns of the secondary coil but also the iron core itself, for instance, in a transformer. If this happens, energy will be lost from the secondary circuit and efficiency reduced. Figure 10.11 shows a

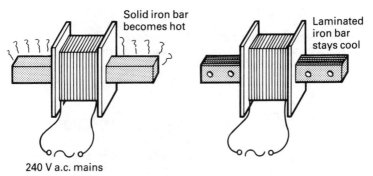

Fig. 10.11 Use of a laminated bar to minimise eddy current heating

simple way of minimising this effect in transformers where these *eddy currents*, which circulate because of the induced e.m.f. in the iron, produce heat in the core.

Making the iron core of insulated layers rather than a solid block maintains its magnetic properties but prevents eddy currents from circulating because of the high resistance between layers. Transformer cores are made of metal 'stampings' (Figure 10.9), and the core of an induction coil (Figure 10.3) would be a bunch of thin iron wires rather than a simple rod.

10.11   Uses of eddy currents

Galvanometer damping   A moving-coil galvanometer (Figure 9.14) has its coil wound on an aluminium frame rather than a plastic one in order to take advantage of eddy currents. The frame acts as a one-turn coil and, as it moves when the instrument is used, an e.m.f. is induced in it. The e.m.f. causes a current to pass around the frame which is in such a direction as to oppose the movement of the coil (Lenz's Law, section 10.5). The coil is *damped* in this way and the needle moves steadily to its final position without overshooting or oscillating. The eddy current in the frame exists only whilst there is movement and does not affect the final stationary deflection.

Electromagnetic braking   Figure 10.12 shows the principle of *electromagnetic braking*. Electric locomotives can make use of this effect when travelling down steep gradients. The braking increases in strength the faster the train goes – just the right property required.

A well mounted aluminium disc spins freely by itself . . .   . . . but will not spin more than half a turn in a magnetic field.

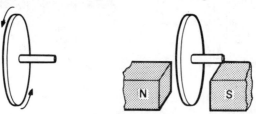

Fig. 10.12   The principle of electromagnetic braking

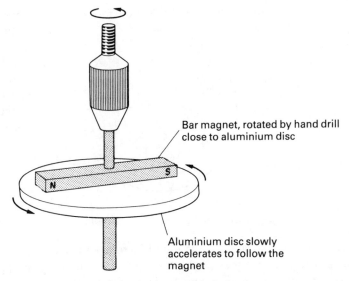

Fig. 10.13 The principle of the induction motor

Induction motors The idea behind *a.c. induction motors*, which are easily the most common type used in engineering, is illustrated in Figure 10.13. The moving magnetic field induces an e.m.f. in the disc, which causes eddy currents, which oppose the motion of the field (i.e. the disc tries to catch up with the field, or is dragged round by it). Figure 10.14 shows a simple application of the effect in a car speedometer.

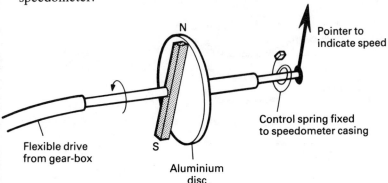

Fig. 10.14 Use of the induction effect in a car speedometer

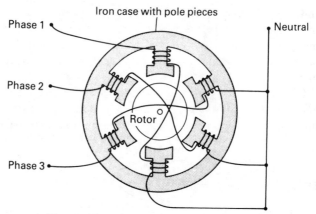

Fig. 10.15 A 3-phase induction motor

A more sophisticated use is found in rotary induction motors where a rotating magnetic field drags round a centrally mounted rotor (Figure 10.15). The moving field is generated by using a single-phase a.c. supply to two sets of pole pieces with a capacitor to shift the phase of one set, or, more usually for heavy work, taking a three-phase supply to three sets of poles. Chapters 11 and 13 deal with these points in more detail.

The main advantage of induction motors is their absence of electrical contacts to moving parts.

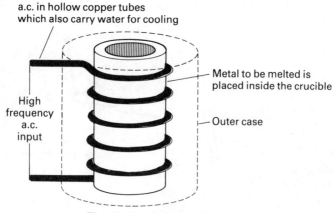

Fig. 10.16 An induction furnace

Induction furnace Eddy currents can be used to heat metals to their melting points, an effect used in the *induction furnace*. The metal is held in a fireproof crucible and a large alternating current passing through the copper tubes induces eddy currents in the metal. Figure 10.16 shows the basic design.

Considerable amounts of power are needed for this to work well and cooling is necessary through water pumped along the copper tubes.

10.12   Self-induction

In any electromagnetic device where there is a changing field, an e.m.f. will be induced in any conductors nearby (as was mentioned in section 10.10), including the primary windings of an induction coil or transformer. Figure 10.17 shows a simple circuit in which this *self induction* is present.

When the circuit is switched on, the current rises slowly due to the 'back' e.m.f. induced in the coil which opposes the growth of current. On switching off, though, the neon lamp will flash. This is due to the rapid collapse of the magnetic field and the high induced e.m.f. because of it. The neon lamp might require 200 V for it to flash. Without the lamp there would be a large spark at the switch contacts when the circuit is broken – and a nasty shock if someone's hand accidentally touched them.

One way of avoiding the high induced e.m.f. is to connect a large capacitor across the coil (Figure 10.3 shows one in the case of an

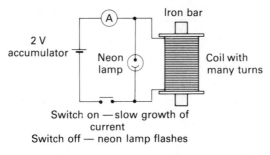

Fig. 10.17   Demonstration of self-induction

induction coil). Any induction device might need protection of this kind against sparks when switching off. (See Chapters 16 and 17 for more about inductive circuits.)

10.13   Back e.m.f. in d.c. motors

In Chapter 9 the principles of d.c. motors were explained in some detail, but the self-induction effect in the armature coils leads to another important consideration. (The iron armature needs to be laminated of course, like a transformer core, to reduce eddy currents in it.) The faster the motor goes, the greater will be the e.m.f. induced in the coils themselves and this will act against the e.m.f. driving the motor. A motor designed to work off a 50 V supply, for instance, can have a *back e.m.f.* as large as 47 V at its operating speed. With an armature coil resistance of 1 Ω, the effective p.d. of 3 V will mean a current through the coils of 3 A.

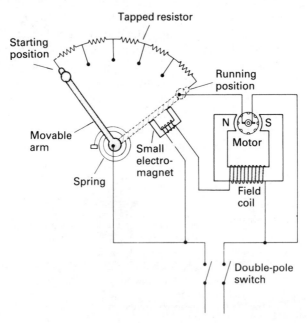

Fig. 10.18   Use of a starting resistor with a d.c. motor

However, when first switched on, the motor starts from rest and the back e.m.f. will not be present, so that almost all the 50 V can apply to the armature coils, giving a current of up to 50 A!

What is needed, to overcome this problem, is a resistor which is progressively removed as the speed of the motor builds up. Figure 10.18 shows one way of achieving this.

Note the use of an electromagnet to hold the arm in the running position, and of a spring which returns the arm to the starting position when the motor is switched off. (This type of electro-mechanical control circuit is intended to illustrate the principle of using a resistor that can be phased out as the motor gains speed. The same effect can be achieved by electronic control systems which have no moving parts and which require no maintenance.)

10.14   Summary

Movement between a conductor carrying a current and a magnetic field induces an e.m.f.

Changes of the current's size or direction induce an e.m.f.

The direction of the induced e.m..f is at right angles to both the magnetic field and the movement.

The induced current, if any, is such as to oppose the change.

Induction coils use a mechanical device to produce a changing current.

Transformers operate only with a.c.

For a transformer,

$$\frac{V_{in}}{V_{out}} = \frac{\text{number of turns on the primary coil}}{\text{number of turns on the secondary coil}}$$

For a perfect transformer, $I_{in}V_{in} = I_{out}V_{out}$

Energy losses in transformers arise through heating, eddy currents and magnetisation in the core.

The high efficiency of transformers leads to the use of a.c. for the distribution of electrical energy.

Eddy currents occur in any conductor near to a changing magnetic field.

Large scale a.c. applications use induction motors.

Self-induction can cause sparking.

A back e.m.f. is generated in d.c. motors.

11

Generating a.c. and d.c.

11.1 The size of the problem

The existence of a reliable and safe system which enables domestic, commercial and industrial users to plug into an electricity supply is taken for granted in developed countries and demanded in many developing countries.

The needs of consumers vary from small households to large manufacturing complexes, scattered across the country. Each requires its own supply, appropriate for its demands and sometimes at a particular voltage, independent of other users wherever possible, and available twenty-four hours a day. In addition, there is a preference for the distribution cables to be out of sight, and for an interconnected system which does not fail when the local power station suffers a breakdown. Very high peak demands must be met depending on the weather or time of day or special events. Standardisation of provision is needed across the country yet the large conurbations and industrial areas might have to be given priority status during emergencies or particularly severe circumstances.

All this points towards a major industry and a commercial organisation of large proportions, employing tens of thousands of people, needing raw materials in massive quantities and responsible for the distribution of thousands of millions of watts of electrical power.

11.2 Generators and dynamos

The induced e.m.f. produced from a magnetic field and movement of a wire through it gives the basis of a generator or dynamo. A

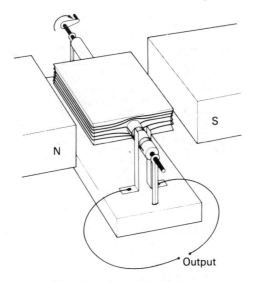

Fig. 11.1 A simple generator

simple generator is shown in Figure 11.1; compare this with Figure 9.7.

The construction of a d.c. generator is identical with that of a d.c. motor, commutator and all, except that the generator is turned mechanically to produce electricity whereas the motor is driven by a power supply to produce motion. Figure 11.2 shows the coil in two positions as it rotates. In (*a*) the magnetic flux passing through the coil is changing at its fastest (though is actually zero at the instant shown) because the sides of the coil are cutting through the field

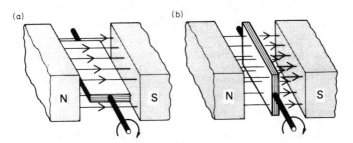

Fig. 11.2 Coil positions for (*a*) maximum and (*b*) zero induced e.m.f.

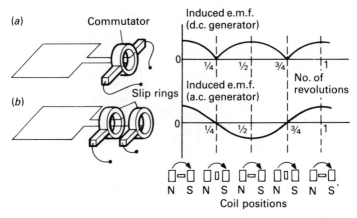

Fig. 11.3 Outputs from single coil generator for (*a*) d.c. and (*b*) a.c. operation

lines at the greatest rate. In (*b*), however, the flux through the coil is largest but is not changing at all when the coil is at this point because no lines of flux are being cut by the coil. At (*a*) the induced e.m.f. will be at its largest, but at (*b*) it will be zero. The e.m.f. across the ends of the coil would therefore be a fluctuating one which is sketched graphically in Figure 11.3 for both d.c. and a.c. operation.

11.3   Practical generators

The design of real generators is more complex than the simple laboratory model and includes features such as concave pole-pieces, an electromagnet in place of the permanent magnet, multiple coils of wire connected in series wound on a laminated iron core, and carbon brushes for contacts to the commutator (Figure 11.4). These are very similar to the practical details of electric motors (section 9.5) and often the same device can be used either as a motor or as a generator.

11.4   The cycle dynamo

A common kind of generator is that fitted to bicycles to provide power for the lamps, often driven by a knurled pulley that bears on

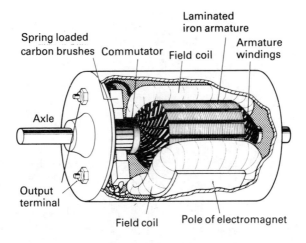

Spring loaded carbon brushes Commutator Field coil

Laminated iron armature

Armature windings

Axle

Output terminal

Field coil Pole of electromagnet

Fig. 11.4 Details of a practical generator

one of the cycle's tyres as illustrated in Figure 11.5(*a*). Here a permanent magnet is used and the coil is wound on an iron armature. Slip ring contacts are fitted to lead to outside connections, one of which is the case of the dynamo – i.e. the metal frame of the bicycle. (This is why only one wire is used to connect dynamo to lamp, the frame providing the other conductor.)

The output of this type of dynamo will be a.c. but not as smoothly varying as the mains supply, because there will be high and low spots at certain points of the armature's rotation. Figure 11.5(*b*) shows a typical e.m.f.-time graph for a dynamo output which could be displayed on a CRO (section 6.8).

11.5 Power station generators

Very large generators which produce tens of megawatts of power are designed on a different principle from those described above. Instead of the magnetic field being provided from a stationary magnet or electromagnet and the coils in which the e.m.f. is induced rotating within the field, the opposite arrangement is used: the

(a)

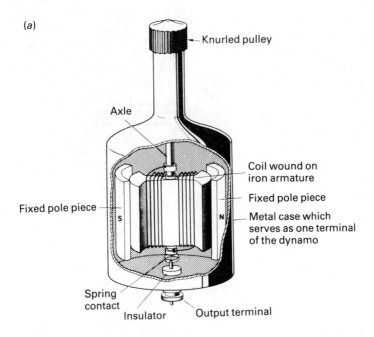

(b) ·

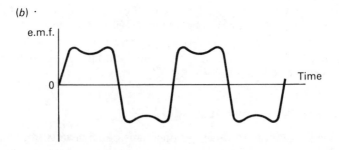

Fig. 11.5 Construction and typical output of a bicycle dynamo

magnetic field rotates and the coils remain stationary. Figure 11.6 shows a cross-section through such a generator. The rotor is driven by steam turbines in the power station and carries its own magnetising coils fed by a separate d.c. supply. Each pair of 'arms' of the rotor is magnetised N–S alternately, so that N and S poles pass close to each coil of the stator in turn. The diagram is of a *single phase* alternator, all the stator coils being joined in series to give a single output.

A more complex arrangement uses three different outputs from opposite pairs of coils, which produces three a.c. voltages out of phase with one another (section 13.7). Figure 11.7 (p. 126) shows a schematic layout of the turbine which drives the generator.

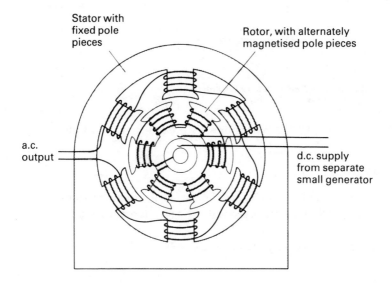

Fig. 11.6   Single phase a.c. generator

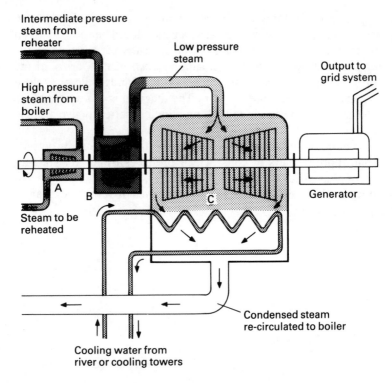

Fig. 11.7 Main features of a power station turbine unit

A high pressure turbine
B intermediate pressure turbine
C low pressure turbine

11.6 Summary

Generators of electricity are similar in construction to motors.

In large a.c. generators the magnetic field is made to rotate close to fixed coils.

a.c. generators can produce single or three-phase outputs.

12

Changing a.c. to d.c.

12.1 The nature of the problem

Many appliances such as tape recorders, calculators, radios and television receivers operate on direct currents and voltages, but need to be powered by the a.c. mains supply. The requirement is therefore to change the a.c. to d.c. and in most cases, to reduce the voltage to a much lower vlaue.

Transformers can easily change the voltage level of the a.c., of course, but to convert the a.c. to d.c. requires a one-way device which can simply eliminate one half of the a.c., together with some way of smoothing out the remaining ups and downs to a single steady value. The whole process is called *rectification*.

12.2 Half-wave rectification

A *diode* is a semiconductor device which conducts an electric current much more readily in one direction than the other (sections 4.3 and 19.3). Most diodes are quite small, about 1 cm long and 2 mm or 3 mm wide, totally sealed into cases with two contact wires emerging for connecting into circuits (Figure 12.1).

The typical voltage-current conduction curve for a diode shows a low resistance in the forward direction but a much higher one in the reverse direction (Figure 12.2). Note the different scales on the graph which indicate a forward resistance of perhaps $10\ \Omega$ but a reverse resistance of $100\ M\Omega$.

A simple circuit containing a diode can show how it conducts easily in one direction but hardly at all in the other (Figure 12.3).

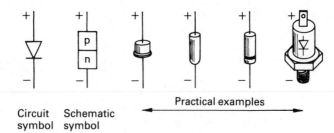

Circuit   Schematic |← Practical examples →|
symbol   symbol

Fig. 12.1   Types of diodes and circuit symbols

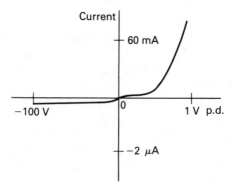

Fig. 12.2   Conduction curve for a diode

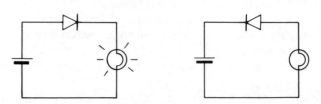

Fig. 12.3   The diode as a one-way conductor

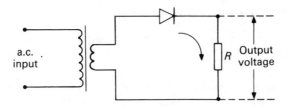

Fig. 12.4 A simple rectifier circuit

The simplest rectifier circuit consists of a step-down transformer, a resistor and a diode (Figure 12.4) connected to a 'load' of resistance *R*, such as a d.c. motor.

The input and output voltages are indicated in Figure 12.5 as a CRO would display them. The p.d. across the resistor is present only when the diode conducts in its forward direction, the negative half of the original a.c. being almost entirely suppressed. The output voltage is certainly above the zero line, but it is far from steady; it can be regarded as a mixture of a.c. and d.c. or as a fluctuating d.c.

Such a basic rectifier could be used as a battery charger or for driving a model car or train where the only requirement is for a unidirectional voltage. It does, however, waste half of the a.c. input by simply blocking it out. The name, *half-wave* rectifier, indicates this effect.

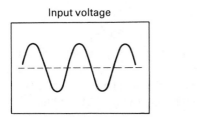
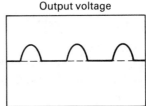

Fig. 12.5 Half-wave rectification

12.3   Half-wave rectification with capacitor

A much smoother output, even with a simple half-wave rectifier, can be obtained by adding a *reservoir capacitor* in parallel with the resistor (Figure 12.6).

When the diode conducts, the capacitor is charged up and after a few cycles reaches a p.d. nearly at the peak of the a.c. On the negative peaks when the diode does not conduct, the p.d. across the capacitor sends a current through the resistor which gradually falls until the next conducting period tops it up again. The capacitor acts like a reservoir which is filled intermittently but also has a continuous leak from it, the level in the reservoir going up and down by only a small amount. The p.d. across capacitor and resistor is much steadier than the simple half-wave case. What matters is that the rate of leaking is much slower than the rate of topping up (see also Chapter 15).

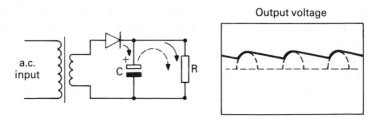

⟶   Diode conducts through R and charges up C
--⟶   Diode off but p.d. across C provides some current rhrough R

Fig. 12.6   Half-wave rectifier with capacitor

12.4   Full-wave rectification

To take advantage of the whole of the a.c. input it would obviously be better if the negative part of the cycle could be used, as well as the positive half. In particular, if the negative half could be turned over rather than blocked out that would fill in the gaps of the half-wave circuit.

There are two ways of doing this (Figure 12.7), one using two

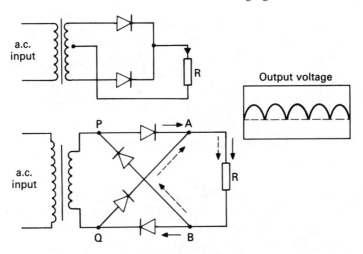

Fig. 12.7 Full-wave rectifier circuits

diodes with a centre tapped transformer, the other using four diodes. Each has the same effect of making current pass through the resistor in the same direction all the time. In the latter case, which is more commonly used, the current passes alternately along PABQ and QABP.

It is possible to buy the four-diode network as a single unit with labelled contacts for a.c. in and d.c. + − out.

12.5 Full-wave rectification with capacitor

Adding a reservoir capacitor to a full-wave circuit has the same topping-up effect, but as it occurs twice in every a.c. cycle the rectified voltage (or current) is steadier than the half-wave case (Figure 12.8).

12.6 Removing the ripple

Even with full-wave rectification and a reservoir capacitor the output is still not quite steady – there is a 'ripple' on the d.c. level.

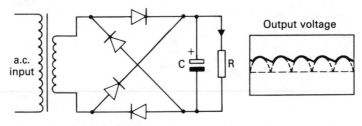

Fig. 12.8   Full-wave rectifier with capacitor

For many purposes this does not matter, but for radio or TV or amplifier use the small changes in the rectified supply would be magnified and come through as a continuous buzzing sound of 50 Hz or 100 Hz (half-wave or full-wave).

An arrangement which removes the ripple almost entirely is shown in Figure 12.9, using two reservoir capacitors and an *inductor* or *choke* (see Chapter 16). The inductor reacts against any change of current (section 10.5) and helps to keep the output almost level.

For very high quality applications it is possible to use more elaborate filter circuits which remove the remaining ripple to such an extent that it cannot be noticed.

Note the way this last circuit is drawn with components spread out and connected individually to the + and − lines. In practice, of course, common connecting points might be used wherever possible or printed circuit boards with some connections already made would have the various units mounted on them directly.

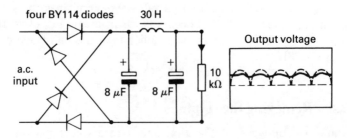

Fig. 12.9   Full-wave rectifier with smoothing circuit

Some typical values of components are also shown in Figure 12.8, the exact choice depending on the intended use and the current required from the rectifier. The voltage would be determined by the choice of transformer.

The cathode ray oscilloscope is a good example of the use of rectified a.c. The electron gun operates off a few volts a.c. for its heating filament, but cathode, grid and anodes require d.c. voltages which vary from -30 V to $+2000$ V. A single transformer with several separate windings, both step-up and step-down, with appropriate rectifier circuits would be used to drive the whole instrument.

12.7 Battery eliminators

With the development of portable and transistorised equipment of many kinds has come the use of dry cells to power them. Often the voltage needed is between 3 V and 9 V, and the currents not very large, so the dry batteries provide an adequate source of electricity. As an optional extra with a portable TV, tape recorder or even a calculator many manufacturers offer a mains connector or *battery eliminator*. Such a device is, of course, simply a transformer rectifier packed into a small case and designed to deliver the required voltage (d.c.) through a special plug connector.

12.8 Summary

Rectification is the conversion of a.c. to d.c.
Half-wave rectification requires a single diode.
Full-wave rectification requires two or four diodes.
A rectified a.c. can be smoothed by using a reservoir capacitor.
Ripple on a rectified voltage may be reduced by means of a choke.

13

Distributing and Using Electricity

13.1   The UK National Grid distribution system

Electricity is distributed throughout the United Kingdom by means of the National Grid system, which is an interlinked network of power stations and cables connecting houses and factories to the generators themselves. Electrical power is generated at 25 kV but immediately stepped up to 275 kV or 400 kV for distribution. Depending on the needs of the consumer, it is then stepped down at transformer sub-stations to the appropriate voltages, the lowest being for domestic use (Figure 13.1).

The total capacity of the grid system is designed to meet the heaviest demands which come at peak times during the day and night or in very cold winters. Mostly there is spare capacity which enables maintenance and repair to be carried out with some power stations temporarily not contributing to the system. A Control Centre monitors the demand and can if necessary switch in additional supplies or transfer them from one area to another. Complete breakdowns in supply are therefore rare.

13.2   Why a.c. is used rather than d.c.

At first sight it seems unnecessarily complicated to use a.c. instead of d.c. for the electricity supply, since in many appliances the first thing to be done is to change it to d.c.! The reason is one of cost in transmitting electricity over large distances.

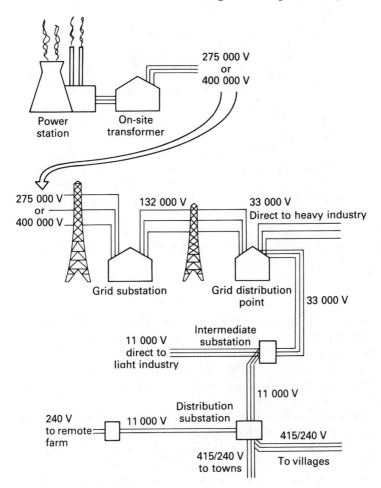

Fig. 13.1 The distribution network for electricity

Figure 13.2 shows two cases where a power station has to deliver 1000 kW, one using 10 kV at 100 A, the other 100 kV at 10A, through cables which have a resistance of 10 Ω. In the first case, the heat transfer in the cables (I^2R (section 8.2)) is $100^2 \times 10$ W = 100 000 W or 100 kW, leaving only 900 kW of power for the

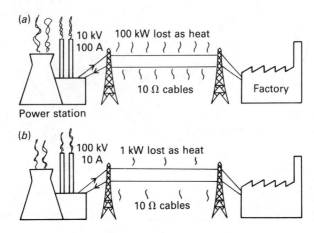

Fig. 13.2 Higher voltage transmission leads to lower energy loss

consumer. In the second case the heat transfer is only $10^2 \times 10 =$ 1000 W or 1 kW, leaving 999 kW of useful power. The lower current leads to a smaller heat 'loss' but it needs a higher voltage. It is therefore advantageous to use high voltage for transmitting electricity over large distances. In addition, thinner cables can be used for the lower currents.

However, very high voltages can be extremely dangerous and would not be suitable for domestic or industrial use. The ability of transformers to step-up or step-down a.c. voltages (section 10.7) provides the link which enables different voltages to be used without significant power losses. Because there is no comparable way of transforming direct voltages, it follows that high voltage a.c. transmission is the chosen system.

A small scale laboratory model to illustrate the advantage of high voltage transmission is illustrated in Figure 13.3. In (a) a 12 V a.c. source feeds two lamps, one directly and the other at the end of wires with a resistance of a few ohms; the 'distant' lamp glows only very dimly. In (b) the voltage is stepped up to, say, 240 V and down to 12 V at either end of the transmission wires; this time the distant lamp is at almost full brightness.

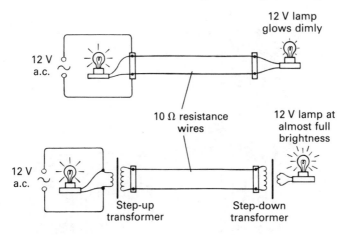

Fig. 13.3   Demonstrating the benefit of high voltage transmission

13.3   Electricity and the home

From the National Grid, electrical energy is supplied to domestic users often through small local transformers, such as those shown in Figure 13.4, which step down the level from 415 V to the standard British voltage of 240 V. (Other countries have different values, e.g. 115 V, for use in homes.) The final connections are made through underground or overhead cables. Two wires enter the house, labelled L (for live or line) and N (neutral) as shown in Figure 13.5. A third wire, E (earth), is often found connected to a metallic object in the ground, giving it a zero potential.

Notice the fuses (section 13.4) fitted into the live wires. The distribution box can serve several circuits such as the one illustrated, which is a *ring* circuit with a spur. Ring circuits are used for power points, which can be connected anywhere in the ring, and have the advantage of needing less cable than if each point were supplied separately. A medium sized house might have two ring circuits, one for upstairs and one for downstairs. Lighting sockets need far less current than power points and are often connected through a simple parallel arrangement. Special circuits which carry large currents, for an immersion heater or electric cooker, for example, have their own separate cables.

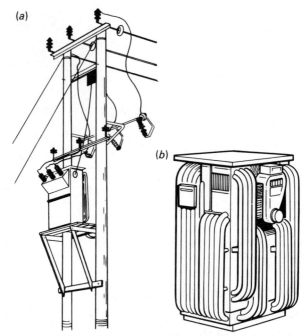

Fig. 13.4   Two transformers in the Grid system used (*a*) to supply an isolated farm, and (*b*) for a village or town district

13.4   The use of fuses

Virtually all electrical circuits, from the largest in the National Grid to those in portable electronic equipment, need to be protected either by *fuses* or *circuit breakers*. Electricity is a great boon to mankind but can also be extremely dangerous. Even a moderate amount of overheating can lead to damaging fires or poisonous fumes being produced, so it is vital that no such risk is taken.

A fuse is simply a piece of thin wire or metal which melts when a current greater than a certain value passes through it. As part of a circuit, the fuse will melt and stop the current.

Figure 13.5 shows that there is a master fuse belonging to the Electricity Board which would disconnect the whole house from the external mains supply in the event of a serious overload. This might be a 60 A fuse. Each circuit connected at the distribution box has its own fuse, ranging from 5 A to 30 A, which is matched to the maximum permitted safe current for the cables in that circuit. Figure 13.6 shows the most common types of fuse.

The fuse with a bare wire has the advantage that it is easy to see whether or not it has 'blown', but the disadvantage of needing a screwdriver (and some patience, too) to replace it. The cartridge fuse, on the other hand, is easy to replace but not always easy to check even if the wire itself is in a glass tube.

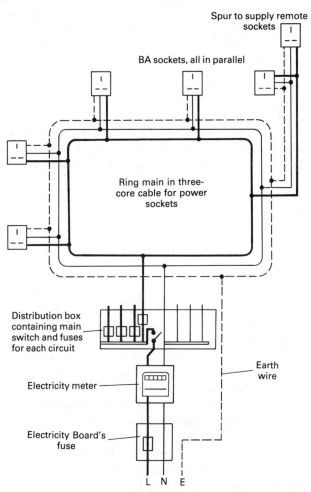

Fig. 13.5 A house wiring system

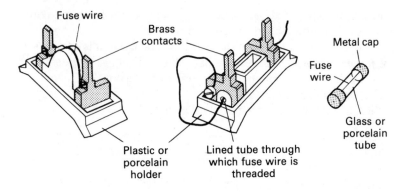

Fig. 13.6   Types of fuse

It is convenient to have all the fuses brought to a common point with some spare wire or cartridges to hand. Motor cars as well as houses have fuse boxes with all the circuits labelled and spare fuses supplied. It is often surprising that car circuits take such large currents (main fuse 55 A, circuits 10 A or 15 A), but with only a 12 V (or even 6 V) system these are needed to obtain the necessary wattage.

13.5   Circuit breakers

A modern alternative to the fuse is the circuit breaker, an electro-magnetically operated switch that 'trips' when a certain current flows through it, thus breaking the circuit, rather like the ejector button fitted to the sockets of electric kettles. A domestic distri-bution box would have perhaps six circuit breakers, one for each of its circuits, each labelled with the tripping current and colour coded. There would also be a master circuit breaker installed with the main switch. The great advantage over fuses is that once the fault has been corrected the circuit breaker can be reset by pressing the control button, without the need to replace fuses. Circuit breakers also respond more quickly and are reliable in operation even when the rated value is only just exceeded.

Circuit breakers are the main safety feature on the National Grid where very much larger currents are involved than fuses could handle.

13.6 Fuse sizes

It is very important that the right size of fuse is selected for each separate appliance so that every appliance is protected individually. If a fault occurs which causes a larger than normal current to flow, it is better for the appliance's own fuse in its plug to 'blow' rather than that in the distribution box. Protection is obtained at the lowest possible current values in this way.

Plugs are often bought with the fuse fitted (as a rule a 13 A one) whatever appliance the plug is for. A range of different fuses for plugs is available – 3 A, 5 A, 10 A, 13 A – and the correct one to choose is a value just greater than the working current taken by the appliance. To find the current, if it is not stamped on the manufacturer's label, it must be worked out from the power (which will be on the label) and the voltage. Some common cases are listed below (see also section 1.5).

Appliance	Power (in watts)	Voltage (in volts)	Working Current (in amps)	Recommended fuse (in amps)
Lamp	100	240	0.42	3 (or smaller if available)
TV receiver	195	240	0.79	3 (" ")
Hair dryer	400	240	1.7	5
Iron	750	240	3.1	5
Toaster	1500	240	6.3	10
Fire	2000	240	8.3	10
Kettle	2750	240	11.5	13

Note the figure recommended for a hair dryer, which is substantially greater than its working current. The reason for this is the starting-off current taken by the motor when first switched on, which can be much higher than its operating current (section 10.13).

13.7 Three-phase a.c.

In section 11.5 it was stated that power stations actually generate electricity from three-phase alternators (Figure 10.15) which produce three outputs of similar sizes but out of step with one another. The a.c. voltages from the three pairs of coils are sketched in Figure 13.8.

All three voltages have the same amplitude and frequency, but the maximum e.m.f. is reached in coil 2 slightly later than coil 1, and in coil 3 later than coil 2. They are said to be *out of phase* with one another by 120°, taking 360° as a complete rotation or cycle from the alternator, since the three curves are equally spaced within the cycle (see also Chapter 17).

Each phase is transformed and distributed over the grid system separately, which is why there are often three or six sets of cables suspended from transmission towers or pylons. There is also a single cable joining the tops of towers without insulation, called the neutral conductor which is connected to earth at the power station.

Industry uses all three phases for much of its equipment which is designed not to need the netural. For domestic purposes, however, two wires must be provided into each house, connected to one of the

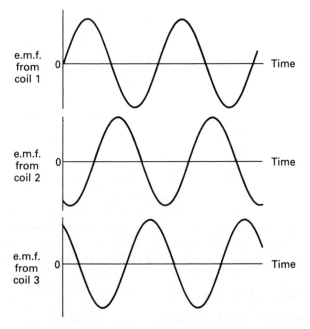

Fig. 13.7   Output from a 3-phase generator

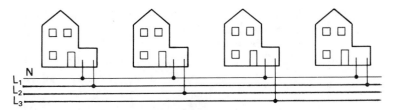

Fig. 13.8   Use of three-phase supply to adjacent houses

phase lines and to the neutral (Figure 13.9). The advantage of this arrangement is that if each phase is connected in sequence to adjacent houses the total current to be carried by the neutral is very small when nearly equal currents are required from each phase. (At any particular time on Figure 13.8 the total of the three graphs is zero!) Thus the neutral cable can be much thinner (and cheaper) than the other cables and the costs of distributing electricity are lower this way. A single-phase system would need two heavy cables to each house multiplied three times to carry the same power.

13.8   Electrical safety – plugs and sockets

The wiring of plugs and sockets is not difficult, but there are some important safety points which it is easy to overlook. Figure 13.10 shows a typical 13 A mains plug with points of good practice indicated by the labels. Similar precautions apply to the wiring of lamp sockets and other items where connections have to be made manually.

13.9   Electrical safety – earthing

Appliances which are wholly contained in plastic casings, with no metal parts accessible to touch, are often provided with twin cable and no earth connections and are perfectly safe. Food mixers, grinders, vacuum cleaners, tape recorders and many other appliances are made in this way and do not need earthing. Metallic casing on appliances, which people can touch or which are used for carrying them around, should be earthed to avoid shocks and possibly deaths. This is done by the earth wire being connected

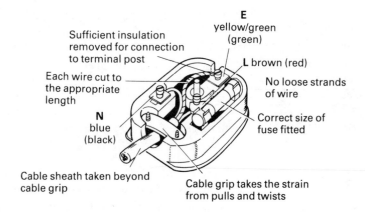

Fig. 13.9 Safety features when wiring a mains plug

through the plug to the casing of the appliance. If by chance the live wire became disconnected and touched the case, a large current would pass between live and earth, causing the fuse in the plug to blow. This would immediately stop the current and the case would be safe to touch. In the absence of the earth connection the case would remain live through the faulty contact and a person who touched the case would receive a nasty shock, perhaps even a fatal one.

13.10 Electrical safety – overloading

A fuse of the right value will protect an individual appliance without putting others out of use, but it is just as important that the cables which feed any socket should not carry unsafe currents. Using a lamp connector for an electric fire, for example, would run the risk of causing a fire in the lighting circuit wires by overheating them. Since lamps take no more than 1 A, the wires used for them might have a 5 A rating at the most, but an electric fire could take 8 A or 10 A. Such a large current would immediately blow the light circuit's fuse of 5 A. If a larger fuse had been inserted in error, however, the wires in the lighting circuit would be overloaded.

Another common cause of overloading occurs with the use of adaptors which can allow two or three separate plugs to be used off one socket. Some adaptors have their own fuses, say 13 A, for their own protection, so the total current used could not exceed that value. Many adaptors, however, have no fuses and the danger occurs when the three appliances together take more than the safe current for the cables connected to the socket. If a plug or socket becomes hot in use always check the wiring of it and the total load it is carrying. It is heating devices which can most easily cause overloading, because of the high currents they require.

13.11 Electrical safety – preventing shocks

A shock from the 240 V mains is an unpleasant and frightening experience. It is the current through the body, especially the chest, which causes the damage and, in some cases, death. The current depends on the voltage, of course, but also on the resistance of the 'circuit' through the body to the earth or from one hand to the other. Small voltages, less than about 50 V, do not lead to currents large enough to be dangerous or perhaps even felt, but anything over 100 V certainly can. Even the European and American standard of 115 V is not to be treated lightly and Britain's 240 V can be very dangerous indeed.

When dealing with electrical equipment the best thing is to switch the mains off, of course, and to pull the plug out of the socket. With live equipment, it is a sensible precaution to wear good insulating shoes and gloves. A small 'mains-tester' screwdriver is useful too, to show which contacts are live.

Fatal accidents still occur in homes where unnecessary risks are taken with electricity; the following precautions should be observed:

— do not use mobile electric heaters in bathrooms;
— switch off a light unit before replacing a lighting filament;
— never drill a wall directly above or below a mains socket or light switch unless you *know* which way the cables run from it;
— always connect the earth wire to metal-cased appliances;
— do not remove the rear panel of a television set to adjust it whilst still connected to the mains;
— do not overload circuits by using adaptors.

Electric shocks vary greatly in their effects on people. It is always best to take no risks at all (see 'Safety and Practical Work' on pp. 1–2).

13.12   Electrical safety – maintenance

Routine checking of electrical equipment, particularly cables, plugs and sockets, is a good safety habit. Loose connections should be examined and made secure, and frayed cables replaced. Spare fuses of appropriate sizes should be available and any item that becomes unusually hot should be switched off at once and examined. Even the normal flexing of a cable, to an electric iron or shaver, for example, can affect the insulation after some time and lead to difficulties.

13.13   Summary

Transmission at high voltages reduce energy loss in the cables.
Voltages are transformed up and down at various points in the grid system.
Three-phase distribution is more economical than single phase.
E = earth, N = neutral, L = live (or line).
Fuses and circuit breakers are used to protect appliances and cables and to prevent electric shocks.
Earth connections are used for metal-cased appliances.
Correct sized fuses are needed to prevent overloading.
Routine checking of plugs and cables is good safety practice.

14

Static Electricity

14.1 Positive and negative

As long ago as 300 BC it was known that certain materials could be 'charged' by rubbing them with fur or cloth. Resins (like amber) and glass-like substances were used then, but modern plastics can illustrate the effects much more easily. The Greek word for amber – *elektron* – emphasises the link with electricity. Around AD 1600 it was realised that two kinds of 'charge' were possible, depending on the substances used, and the names *positive* and *negative* were given to them arbitrarily (a pity they were not the other way around! – section 1.7). The charge appearing on glass when rubbed with silk was dubbed positive, and that on ebonite rubbed with fur negative. Cellulose acetate and polythene give similar but much larger effects when rubbed with almost any insulator.

Simple experiments like those illustrated in Figure 14.1 show that two positive or two negative charges repel one another, whereas a positive and a negative attract. However, it is not a matter of there being concentrations of charge, like the poles of a magnet; here the whole rubbed area of plastic will exhibit attraction or repulsion.

14.2 Charges by friction

With the development of very good insulating materials it has become easy for charges to be generated between two surfaces. This 'static', as it is called, can be quite amusing, mildly inconvenient or decidedly dangerous, depending on the circumstances. Some examples are:

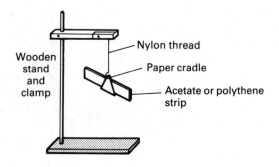

Arrangement before charging

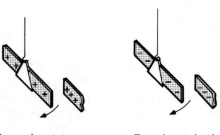

Two charged acetate
strips repel one another

Two charged polythene
strips repel one another

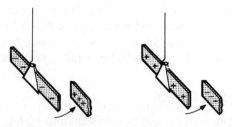

Charged acetate and polythene strips attract one another

Fig. 14.1 Forces between charged materials

(a) the charging of the surface of a record or tape during play which attracts dust particles around the stylus or playing head;

(b) the crackling heard when nylon or other man-made clothes are taken off in dry cold weather;

(c) the electric shock experienced when touching a car door handle after sliding off the seat to alight;

(d) the way adhesive tapes tend to be attracted to their backing paper after having been stripped ready for use;

(e) the disorderly behaviour of newly washed and dried hair after combing;

(f) the charges developed from office carpets which can lead to people feeling slight shocks on touching metal objects;

(g) the sticking of balloons to walls after rubbing them on clothing.

Electrically conducting tyres for cars and aeroplanes have been developed specifically to avoid problems for travellers on disembarking; and in some manufacturing processes where inflammable vapours are produced great care must be taken to avoid even small risks of sparking due to the build up of static.

When materials are rubbed together it is the transfer of electrons which produces the charges on them. In a negatively charged material there are more electrons than usual, but in a positively charged one there is a deficiency of electrons.

14.3 Electric fields

When an object such as a plastic ruler has been charged by rubbing, it can attract small pieces of paper over a distance of a centimetre or two, or cause someone's hair to move towards it. Magnetic attraction (and repulsion) of a similar kind led to the idea of a magnetic field and in the same way we say there is an *electric field* around the charged object.

Field patterns in electricity cannot be as easily demonstrated as in magnetism but they are of a similar kind: they show the direction taken in the field by a small object with equal + and − charges at its ends (called a *dipole*). The lines originate from +ve charges and end on −ve ones, being most closely packed where the field is strongest. Electric fields occur naturally whenever objects become charged. Their strengths are measured in volts per metre. A normal atmospheric field strength would be around 130 V/m, but during thunderstorms the value is very much higher. To obtain a spark between flat electrodes in dry air requires some 30 000 V/cm or 3 MV/m. (Section 14.9 also deals with some aspects of electric fields.)

14.4 Electric potential

Potential differences have been described in Chapter 3 and can be thought of for many purposes as the electrical 'level'. Using the Earth as a zero potential, any charged object might have a positive or negative potential. It is surprising to learn that quite large potentials (or voltages) can be obtained by rubbing materials together – thousands of volts, perhaps – yet they do not pose very great threats. It is similar to the high temperature found in a spark from a flint, for example, which though very hot contains little energy.

14.5 Capacitance

Another important electrical quantity is called *capacitance*. It measures the 'size' of a conductor (roughly analogous to the volume of a container) by the charge required to raise its potential by one unit amount, i.e. volt. A conductor has a large capacitance if it needs a large amount of charge to raise it by one volt, just as a large vessel needs a large quantity of water to fill it to a certain level.

$$\text{Capacitance} = \frac{\text{Charge}}{\text{Potential}}$$

The unit of capacitance is one coulomb per volt, usually called one farad (symbol F) after British physicist Michael Faraday (d. 1867). An object with a capacitance of one *farad* requires a charge of one coulomb to raise its potential by one volt. For most purposes the farad is an impossibly large unit, so the microfarad μF (millionth) is used or even the picofarad pF which is a micro-microfarad (million millionth). Chapter 16 deals with capacitors in more detail.

14.6 Electrostatic induction

A process called *induction* is important with static charges. It occurs when conducting objects are brought close to charged materials and depends on the fact that in conductors some electrons are mobile and will be attracted or repelled through the material of the conductor by external charges. In each case an uncharged conductor has some of its electrons attracted or repelled by the charged object so that an opposite *induced* charge appears nearest to the object. Figure 14.2 shows some simple cases. Note the use of an earth connection to separate the + and − induced charges in the above examples. In practice such a connection can be provided by touching the conductor, the human body being the link to earth.

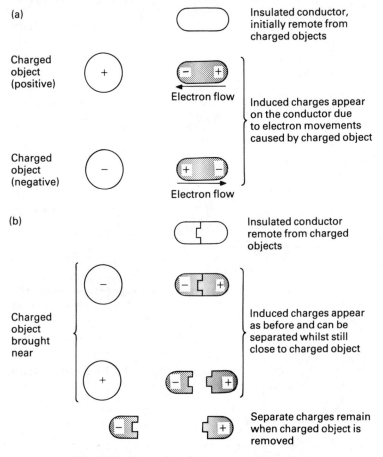

Fig. 14.2 Examples of electrostatic induction

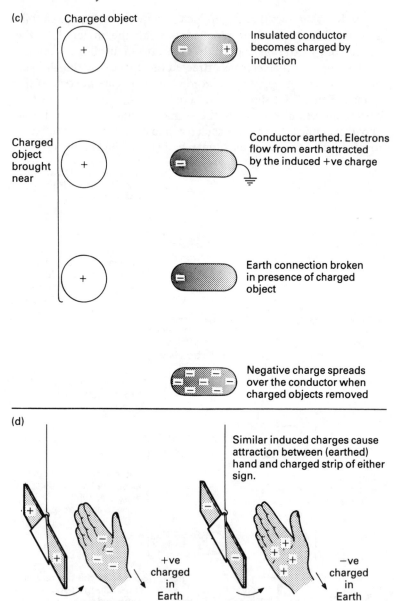

(c)

Charged object

Insulated conductor becomes charged by induction

Charged object brought near

Conductor earthed. Electrons flow from earth attracted by the induced +ve charge

Earth connection broken in presence of charged object

Negative charge spreads over the conductor when charged objects removed

(d)

Similar induced charges cause attraction between (earthed) hand and charged strip of either sign.

+ve charged in Earth

−ve charged in Earth

Fig. 14.2 Examples of electrostatic induction

14.7 Electroscopes and electrometers

Devices which respond to the presence of static electricity are known as *electroscopes* or *electrometers*, depending on whether they give only visual indications (-scopes) or can lead to measurements being made (-meters). Whichever is used, the effects of charges being given to various conductors can be demonstrated and investigated.

The simplest electroscope is made from a thin gold leaf attached to a rod and surrounded by a box or container (Figure 14.3). Deflections of the leaf away from the rod can be viewed directly or by projection through the glass sides of the container.

Electrometers consist of electronic circuits which result in a galvanometer deflection that can be measured against a scale, often with a selection of different ranges (Figure 14.3). Each type of instrument has a metal plate to which the charge is given and which is insulated from the earthed container.

It is tempting to think that these devices indicate electric charge, i.e. measure coulombs, but in fact they are both voltmeters registering the p.d. between the plate (and its attachments) and the container. What a particular charge does is to produce a p.d. depending on the capacitance of the system.

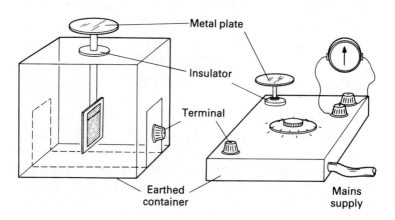

Fig. 14.3 Electroscope and electrometer

14.8   Basic electrostatic effects

Using an electroscope (or electrometer) the following properties of charged objects can be demonstrated. (A standard school textbook should be consulted by readers who wish to see the practical details of how this is done.)

Objects of different capacitances have their potentials raised by different amounts if given equal charges, the higher capacitances acquiring the lower potentials.

The deflection of a charged electroscope is lowered by the approach of an earthed conductor, e.g. a person's hand.

A charged object with a like charge to that of an electroscope increases its deflection.

A hollow charged conductor has all its charge on the outer surface and a zero electric field inside it (section 14.9).

A solid charged conductor has higher concentrations of charge on its more pointed areas (section 14.9).

Sharply pointed conductors can be used either to collect charges or to remove charges (section 14.10).

14.9   Charge distributions on conductors

A charged conductor which is spherical has an even distribution of charges all over its surface, as would be expected, but with other shapes charge is concentrated on the more pointed regions. Figure 14.4 shows the effect for three different shapes.

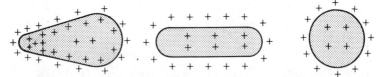

Fig. 14.4   Charge distribution of conductors of varied shape

The most sharply curved areas have the highest concentrations of charge.

With hollow conductors it turns out that there are no charges at all on the inside surfaces, all of it being on the outside, which means that the electric field inside the conductor must be zero. Figure 14.5 illustrates this effect.

A practical application of the hollow-conductor behaviour is found in the use of coaxial cable for connecting between aerial and receiver in TV or car radio sets. Very small signals picked up by the aerial are passed along a wire shielded from unwanted electric fields by an earthed braided conductor (Figure 14.6). Inside the braiding there can be no fields to affect the signal wire.

Another use of the same principle is found in electronic circuits where very sensitive sections are encased in an earthed metal box to screen them from external electric fields.

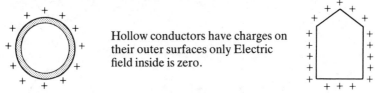

Hollow conductors have charges on their outer surfaces only Electric field inside is zero.

Fig. 14.5   Charges on hollow conductors

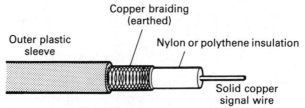

Fig. 14.6   Coaxial cable construction

14.10   Charge effects with pointed objects

Any charged object will eventually lose its charge because of the −ve and +ve ions always present in the surrounding air. If a pointed conductor is attached to such an object the loss of charge is much faster, because of the extra ionisation caused by the concentration of charge around the point (Figure 14.7). Negative ions are attracted to the point and positive ones repelled from it, so the overall effect is that positive charges are 'sprayed' away from the

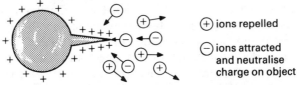

Figure 14.7   The discharging action of a pointed object

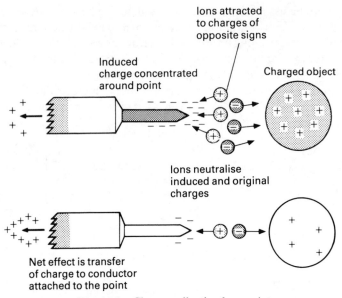

Fig. 14.8   Charge collection by a point

point whilst its own charge is neutralised. A charged electroscope with a needle attached to its plate quickly discharges itself in this way.

By a similar mechanism a point can in effect collect charge by induction from a nearby charged object through the movement of ions between the two of them (Figure 14.8).

14.11   Van de Graaff generator

A device which makes use of both the charge-collecting and charge-spraying action of points is the Van de Graaff generator. Figure 14.9 illustrates its main features. From a moderately high voltage power pack, a much larger voltage can be built up by charge being carried on an endless belt from the pack to a large hollow metal sphere. The size of the voltage will ultimately depend on the smoothness of the sphere and the amount of ionisation or dust in the air. Research machines of this kind with the top sphere surrounded by high pressure filtered nitrogen can be made to develop up to ten million volts. Even small laboratory models will reach hundreds of thousands of volts. A smaller earthed sphere nearby will provide a spectacular spark discharge every few seconds when the voltage builds up to the level at which the air breaks down.

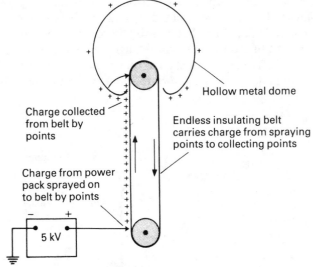

Fig. 14.9 The Van de Graaff generator

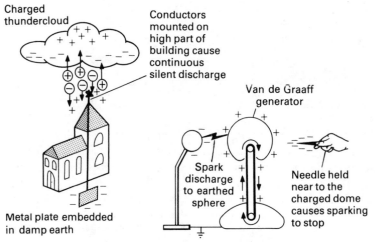

Fig. 14.10 Action of a lightning conductor

14.12   Lightning conductors

Another use of pointed conductors is to reduce the risk of damage caused by a lightning strike from a charged thundercloud. Tall buildings are fitted with a thick strip of metal, usually copper, which is connected to a conductor at the top and a buried earth plate at the bottom. Continuous and harmless discharge occurs via the conductors instead of a disastrous sudden lightning strike (Figure 14.10 (page 157)). The same effect on a small scale can be demonstrated using a Van de Graaff generator to simulate the thundercloud.

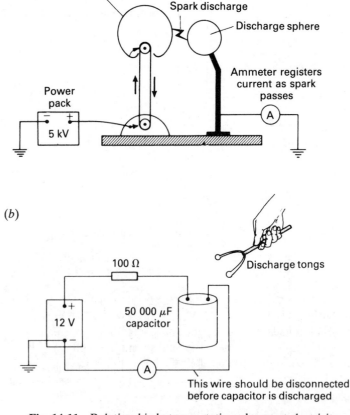

Fig. 14.11   Relationship between static and current electricity

14.13 Static and current electricity

It is important to appreciate that current electricity is merely static electricity in motion. Both are concerned with charged carriers (electrons or ions) and with their effects either on insulated objects or in motion through conductors.

Figure 14.11 shows two arrangements which make the connection between static and current electricity explicit. In Figure 14.11(*a*) the ammeter registers current each time a spark passes from the Van de Graaff generator. In Figure 14.11(*b*) a current is indicated by the ammeter as the capacitor is being charged up. After disconnecting the earthed lead the capacitor can be discharged using the insulated tongs, producing a spark as the contacts are made. In each case the equivalence between an electric current and a movement of charge is demonstrated.

14.14 Summary

Electric charges are + and −.

Like charges repel one another, unlike charges attract.

Objects can be charged by rubbing, which involves a transfer of charges from one material to another.

Electric fields exist around charged objects.

Capacitance is measured in farads.

A farad is a coulomb per volt.

$$C = \frac{Q}{V}, \quad Q = CV, \quad V = \frac{Q}{C}$$

A capacitor can be used to reduce sparking.

Objects can be charged by induction, which often involves an earth contact.

Electroscopes and electrometers are used to indicate the properties of charged objects.

Charges are concentrated around sharp edges and pointed areas of charged conductors, where the field strength is highest.

There is no charge or field inside a hollow conductor.

Pointed objects can lose or collect charges, e.g. in Van de Graaff generators.

Static electricity in motion is current electricity.

15

Capacitance

15.1 Capacitors

The idea of capacitance was met in section 14.5 and defined as the charge required on a conductor to raise its potential by one unit amount, i.e. volt:

$$\text{i.e. capacitance} = \frac{\text{charge}}{\text{potential}}$$

$$\text{or} \quad \boxed{C \text{(farads)} = \frac{Q \text{(coulombs)}}{V \text{(volts)}}}$$

This can also be written as $Q = CV$ and $V = Q/C$. The unit of capacitance, the *farad*, is therefore one *coulomb per volt*.

A device which has capacitance is called a *capacitor* and such things are used in almost all electronic circuits and in many other appliances. The construction of capacitors is quite complex, but in effect they consist of two conducting materials separated by a thin layer of insulation. Any charge given to a capacitor produces a potential difference between the two conductors. Most capacitors can be connected either way round into a circuit, but one type – *electrolytic* capacitors – must be used in accordance with the + and − signs indicated on them. All capacitors have a maximum operating p.d. beyond which there is a risk of electrical breakdown.

Figure 15.1 shows various types of practical capacitors (not drawn to scale), together with the circuit symbols.

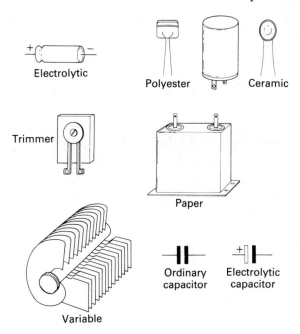

Fig. 15.1 Types of capacitor and symbols

15.2 Capacitors in series and parallel

The formulae for the equivalent capacitance of capacitors connected in series and parallel turn out to be the reverse of those for resistors. (Readers might like to work out each case or to refer to a text book of physics.) The reasons for the difference are that a parallel connection gives a larger capacitance than each single capacitor, whereas a series arrangement leads to a smaller capacitance.

For capacitors of capacitance C_1, C_2, C_3 in parallel, the equivalent capacitance C is given by:

$$C = C_1 + C_2 + C_3$$

For capacitors of capacitance C_1, C_2, C_3 in series, the equivalent capacitance C is given by:

$$\frac{1}{C} = \frac{1}{C_1} + \frac{1}{C_2} + \frac{1}{C_3}$$

15.3 Charging and discharging of capacitors

A capacitor can be charged by connecting it to a d.c. power pack. Electrons are supplied from the negative terminal of the pack to one of the capacitor's 'plates' and removed by the positive terminal from the other (Figure 15.2).

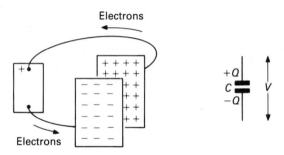

Fig. 15.2 Charges on a charged capacitor

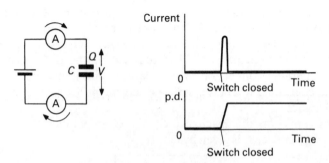

Fig. 15.3 Charging a capacitor

The net effect of this is that equal but opposite charges are produced within the capacitor with a potential difference between them. An ammeter placed in the circuit will indicate a momentary current which stops quickly when the p.d. across the capacitor equals that of the power supply. Figure 15.3 indicates the process.

After being charged up, the capacitor will hold its charge for a long time, though not indefinitely because there is always a small leakage through its insulation or through the moisture of the surrounding air. When its connecting wires are touched together the capacitor discharges and the ammeters indicate a short-lived current in the reverse direction (Figure 15.4).

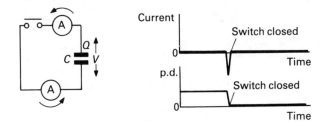

Fig. 15.4   Discharging a capacitor

Note: Capacitors should always be discharged when out of use, to prevent their delivering shocks to unsuspecting people who touch their terminals or try to pick them up.

15.4   Charging and discharging through a resistor – time-constant

In practice it is not possible to charge or discharge a capacitor instantaneously because there is always some resistance present in the circuit, either in the power supply (internal resistance, section 5.5) or in the wires themselves. A truer picture is gained by including a resistor in both charging and discharging circuits (Figure 15.5). The effect of the resistor is to lengthen the charging or

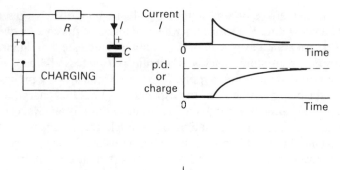

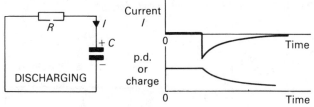

Fig. 15.5   Charging and discharging a capacitor

discharging process over a period of time, as the graphs shows, with the p.d. and charge gradually increasing or decreasing. The current is the rate of flow of charge, and in both cases is high to start with but slowing down after that. (A negative current graph for discharging means it flows in the opposite direction.)

Strictly speaking, the current, p.d. and charge never reach their final values. The shape of the curve is the same as that for a radioactive decay, rising or falling by the same fraction in equal intervals of time. (For the mathematically inclined they are exponential curves of the kind $I \propto e^{-t/RC}$ and $V \propto (1-e^{-t/RC})$.) The main characteristic of this type of curve is the time it takes for a certain proportion of the charge or discharge to be achieved, a quantity similar to the half-life of a radioactive nucleus which indicates the time in which any given level of activity has fallen to half its value. The product RC, called the *time-constant*, represents the time for the charging circuit when sixty-three per cent of the final charge has been attained, or for the discharging circuit when the original charge has fallen to thirty-seven per cent of its value (Figure 15.6).

After only a few time-constant intervals there is very little difference between the values reached and the final values, as the following table shows:

Number of RC intervals	1	2	3	4	5
Rising %	63	86	95	98	99
Falling %	37	14	5	2	1

After three time-constants, ninety-five per cent of the effect has been achieved. For example, if $R = 10$ kΩ and $C = 100$ μF, $RC = 10\,000 \times \dfrac{100}{1\,000\,000} = 1$ second, so after only three seconds just five per cent remains to be achieved and after five seconds a mere one per cent remains.

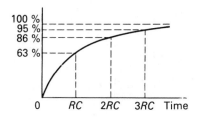

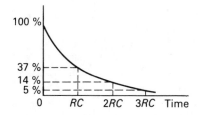

Fig. 15.6 Meaning of time-constant RC

15.5   Capacitors and d.c.

From the previous section it is clear that capacitors cannot conduct steady direct currents: there is no way of passing electrons through a capacitor from one terminal to the other because of the layer of insulation. It is only whilst being charged up or discharged that a current would register on an ammeter, and then only for a short time. For ordinary sizes of resistance and capacitance, say 100 Ω and 10 μF, the time-constant RC is small:

$$\left(100 \times \frac{10}{1\,000\,000} = \frac{1}{1000}\text{ second}\right) \text{ so that a graph like that sketched}$$

in section 15.3 is generally the case.

Thus a 'spike' of current is conducted, but not anything that could be described as d.c. As a variation of this effect, if a d.c. supply could be arranged to switch first one way and then the other (by means of a rotating switch, say), each charging and recharging event would lead to a spike of current (Figure 15.7).

The importance of the time-constant of the circuit can be shown by introducing a resistor into the charging circuit. Depending on the value of RC compared with the repetition time of the switching

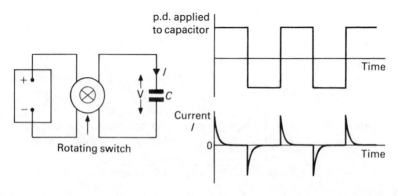

Fig. 15.7   Conduction through a capacitor

device, the current graph will vary as in Figure 15.8. These graphs, which could be displayed on a CRO, should be compared with Figure 12.6, in which a reservoir capacitor is used to smooth out a rectified d.c. voltage. In the latter case a long RC value is required compared with the $\frac{1}{50}$ second repetition time of the a.c. mains.

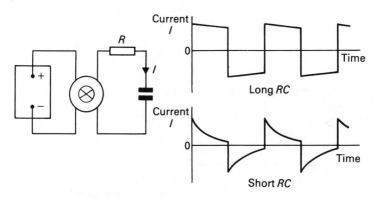

Fig. 15.8 Conduction with different time constants

15.6 Capacitors and a.c.

The mechanically switched d.c. in the previous section is not very different from a.c., which changes direction smoothly rather than suddenly. It follows that some sort of an alternating current will pass under the influence of an alternating voltage, but what is the relationship between the two? A more detailed treatment is given in Chapter 18, but we can say that the current will be greatest when the voltage is changing fastest, and falling towards zero when the voltage is steady, just as in the case of the switched d.c. Figure 15.9 shows the case of a 'pure' capacitance with negligible resistance in the circuit (but a little time after switching on, when the pattern has become established.)

A quick inspection of these curves shows three clear aspects:

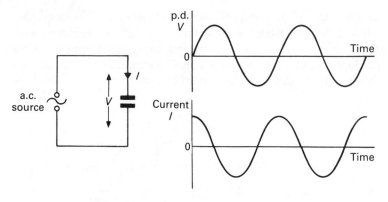

Fig. 15.9   Conduction of a.c. by a capacitor

(*a*)   the current curve is the same shape as the voltage one, i.e. there is conduction of a.c. through the capacitor;

(*b*)   the maximum current has a constant relationship to the maximum voltage, i.e. there is an 'a.c. resistance' operating;

(*c*)   the current curve is 'out of step' with the voltage curve by a quarter of a cycle, the former being proportional to the gradient or slope of the latter.

In Chapter 17 we shall see that the 'a.c. resistance' effect is a frequency-dependent one, which gives an interesting property to capacitors in a.c. circuits.

15.7   Energy stored in a capacitor

When a capacitor has been charged there will be a surplus of electrons on its −ve terminal and a deficiency on its +ve terminal, with a potential difference between them. It therefore has electrical energy stored in it which can do work when it is wholly or partially discharged.

The energy comes from the work done by the supply in charging it up. At first it is easy for charge to be put into a capacitor, since its potential difference will be zero, but as the charging increases more work has to be done to add charge against the increasing p.d. until, in a short period of time, the capacitor becomes fully charged and its p.d. equal to that of the supply. If all its charge Q had been delivered against a constant p.d. V, the work done would be QV (section 3.4), but since it is a case of the p.d. gradually rising from zero to V, the work done is only $\frac{1}{2} QV$.

Energy W stored in a capacitor is therefore $\frac{1}{2} QV$, and will be in joules provided Q is in coulombs and V in volts.

$$\text{Energy stored,} \quad \boxed{W = \tfrac{1}{2} QV}$$

$$\text{or, using } C = \frac{Q}{V}, \quad \boxed{W = \tfrac{1}{2} CV^2}$$

$$\text{or} \quad \boxed{W = \tfrac{1}{2} \frac{Q^2}{C}}$$

When using the expressions involving C, remember that it must be in farads; e.g. a 10 µF capacitor charged to 100 V:

$$\text{energy stored, } W = \tfrac{1}{2} CV^2$$

$$= \tfrac{1}{2} \times \frac{10}{1\,000\,000} \times (100)^2 \text{ J}$$

$$= \frac{1}{20} \text{ J}$$

It is interesting to consider the power involved if this energy is released in, say, 1/100 second by a sudden spark. Power = energy per second, so in this case,

$$P = \frac{1/20}{1/100} = 5 \text{ W}$$

The same capacitor charged to 1000 V would have 100 times the energy, i.e. 5 J, giving 500 W if suddenly discharged. The size of the spark, or the effect on a person unfortunate enough to be holding the capacitor's terminals at the time, would clearly show the difference between the two cases!

15.8   Applications of capacitors

Two uses of capacitors have been described in sections 12.4 and 14.5 – to suppress sparks across contact points and to smooth rectified voltages. Other applications are to be found in electronic amplifier circuits, TV and radio receivers, radar and microwave devices, traffic light control circuits, hearing aids . . . in fact in virtually all transistorised units, electronic organs, oscillators, synthesizers . . . the list is endless. Capacitors, together with resistors, have a very wide range of usefulness in both domestic and industrial appliances. (Chapter 19 will show capacitors used in simple electronic devices.)

15.9   Summary

$$\text{Capacitance} = \frac{\text{charge}}{\text{voltage}}$$

For capacitors in parallel, $C = C_1 + C_2 + C_3$

For capacitors in series, $\dfrac{1}{C} = \dfrac{1}{C_1} + \dfrac{1}{C_2} + \dfrac{1}{C_3}$

When charging or discharging a capacitor through a resistor, the time-constant CR governs the time required to reach a desired level.

Capacitors are used to block direct current.

Capacitors conduct a.c. continuously but with a change of phase.

Energy stored in a charged capacitor is $\frac{1}{2} CV^2$ or $\frac{1}{2} QV$ or $\frac{1}{2} \dfrac{Q^2}{C}$.

16

Inductance

16.1 Self-induction – inductor or choke

The idea of self-induction was introduced in section 10.12. Figure 10.17 showed the e.m.f. caused when a current is switched on or off in a circuit in which magnetic induction is appreciable. A coil of wire wound on an iron core produces these conditions and is said to possess *inductance* and to be an *inductor* (or, an older term, a *choke*).

An inductor is made very much like a single-coil transformer (Figure 16.1) with U- and T-shaped stampings forming an almost closed magnetic 'circuit' and the coil wound on the central stem of the T. The number of turns of wire and the amount of iron in the core determines the size of its inductance. A small air gap is left

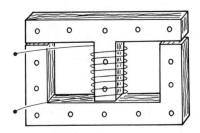

Fig. 16.1 Construction of an inductor

between the bottom of the T and the base of the U laminations, which makes the inductor's effect almost uniform over a reasonable range of current. Even ordinary circuit components and single wires have small inductances but these are generally too low to be significant in practice (except at very high radio frequencies).

16.2 The choke effect

The effect of inductance in a circuit is to produce a *back e.m.f.* whenever the current is altered, i.e. an e.m.f. whose effect is to restrain a rising current but to maintain a falling one. (This is an example of Lenz's Law, section 10.5.) Switching on a highly inductive circuit shows the effect very well by the slow build-up of current (Figure 16.2).

The larger the inductance, the longer it takes for the current to reach its final value, but in each case the final value is the same as that reached if there were no inductance, i.e. a current of E/R. The back e.m.f. simply delays the current growing to its normal size.

Note: In practice the inductor will itself possess resistance which could be included in the value R.

16.3 The henry

Inductance is measured in *henries*, symbol H (after an American scientist Joseph Henry, d. 1878) and its value fixed as the ratio of the back e.m.f. to the corresponding rate of change of current:

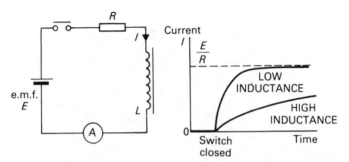

Fig. 16.2 Slow growth of current in an inductor

$$\text{inductance} = \frac{\text{back e.m.f.}}{\text{rate of change of current}}$$

$$1 \text{ henry} \equiv \frac{1 \text{ volt}}{1 \text{ amp per second}}$$

This means that an inductor has an inductance of one henry if a current changing at the rate of one amp per second causes a back e.m.f. of one volt. The faster the current alters, the greater is the back e.m.f. which opposes the rise of the current.

An inductance of 1 H would be quite large, so inductors measured in mH (millihenries) are more common in practice.

16.4   Time-constant in inductive circuits

The rate at which a current grows in an inductive circuit depends on both the inductance and the resistance. The time graph is similar to that met in section 15.4 for the rise of the p.d. across a capacitor (Figure 15.6). Figure 16.3 shows the current curve, and the same way of describing its shape in terms of a *time-constant L/R* as in the case of the capacitor charging-up. (Why the time-constant turns out to be L/R is beyond the scope of this book, but the equation of this curve is $I = \dfrac{E}{R}\left(1 - e^{\dfrac{Rt}{L}}\right)$. Thus for $L = 10$ H and $R = 2\ \Omega$, the time-constant would be $L/R = 5$ seconds, giving a very slow build-up of current, whereas for $L = 10\ \mu\text{H}$ and $R = 1\text{k}\ \Omega$ it would

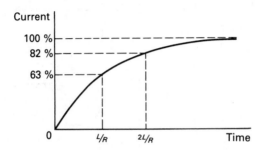

Fig. 16.3   Time constant for inductive circuits

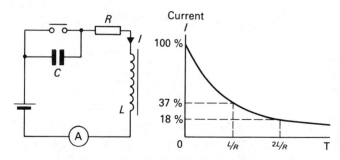

Fig. 16.4   Decay of current when switching off

be $\dfrac{L}{R} = \dfrac{10}{1000 \times 1000}$ second $= 10$ µs and a virtually instantaneous rise straight to the value of E/R.

Switching *off* an inductive circuit produces a high back e.m.f. (sections 10.4, 10.12, 10.13), which can be advantageous or disadvantageous, but this delays the fall of the current with a similar time-constant of L/R (Figure 16.4). (The switch should be protected against sparking by a capacitor.) In this case the e.m.f. acts to keep the current going, i.e. it causes it to fall slowly instead of suddenly.

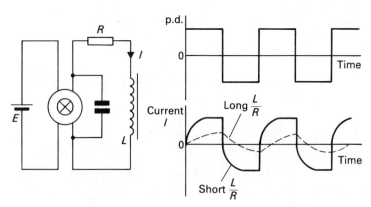

Fig. 16.5   Conduction with different time constants

16.5 Repetitive switching on and off

With a mechanically operated switch (compare Figure 15.8) which could reverse the cell's e.m.f. repeatedly, as in Figure 16.5, the time graphs of current would be as indicated for long and short time-constants. A CRO connected across the resistor R could be used to display these curves. (Why is a capacitor needed in this case?)

The way in which an inductor or choke resists changes in either direction explains its use in smoothing rectified a.c. (section 12.6).

16.6 Inductors and a.c.

Instead of a switched d.c. system for changing the current direction, a.c. can be applied to an inductor to produce a continuously changing effect. Since the inductor is merely a coil of wire wound on an iron former it will conduct electricity easily, but the choke effect means that it will hinder any change of current, trying to prevent the current rising when the applied voltage rises and to keep it flowing when the voltage falls. The result of this is that the changes of current always lag behind the changes of voltage (Figure 16.6). (The current curve in this diagram shows the situation a little time after the a.c. has been switched on, when the pattern has become established.)

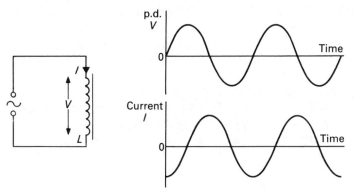

Fig. 16.6 Conduction of a.c. by an inductor

As in the case of the capacitor (section 16.7), these curves show that:

> the current curve is the same shape as the voltage;
> the maximum current is related to the maximum voltage; i.e. there is an 'a.c. resistance' effect;
> the current curve is out of step with the voltage by a quarter of a cycle.

More of the behaviour of alternating currents is discussed in Chapter 17.

16.7   Energy stored in an inductor

As in the cases of a resistor and a capacitor, energy transfers are involved when current flows through an inductor. With direct current it is easy to see that energy is needed to change the magnetic state of the iron core of an inductor. The expression for this works out at $\frac{1}{2} LI^2$ (which compares well with $\frac{1}{2} CV^2$ for a capacitor). Thus a 1 H choke carrying a current of 2 A has $\frac{1}{2} \times 1 \times 4 = 2$ joules of energy stored in it. The important thing to realise is that the amount of energy depends on the square of the current, just as a capacitor's energy depends on the square of its charge or p.d.

16.8   Uses of inductors

In section 14.6 a choke is shown as part of a smoothing circuit for a rectifier. For this purpose a fairly high inductance is needed to give as little rise and fall of current within the repetition time (1/100 second) of the full-wave rectifier as possible. In practice the physical size of the unit might well limit how large an inductor can be accommodated.

Two very important applications of inductors are in oscillators and radio/TV receivers. In both cases the use depends on the behaviour of inductors and capacitors together in circuits using varying currents. Chapter 18 deals with some of these effects.

16.9 Summary

An inductor resembles a single-coil transformer.

Inductance is back e.m.f. per rate of change of current, and is measured in henries.

A henry is a volt per (amp per second).

The time constant for an inductive circuit is L/R.

An inductor chokes a d.c. by slowing down its rise or decay.

An inductor conducts a.c. continuously but with a change of phase.

Energy stored in an inductor carrying a current is $\frac{1}{2}LI^2$.

17

Circuits with a.c.

17.1 Some basic terms in a.c.

When dealing with direct currents and voltages, only two pieces of information are needed to describe them – their sizes and their directions: 5 A or 12 V, together with which way they act, is sufficient to specify the current or the p.d. With *alternating* currents or voltages there are more things that need to be known: amplitude, r.m.s. value, frequency or period, phase and occasionally shape. It is easiest to relate these factors to a time graph of the a.c. in question (Figure 17.1).

amplitude   the maximum value reached in either direction; i.e. so many amps or volts.
r.m.s. value   the root mean square value (section 17.6), the generally quoted figure which is $1/\sqrt{2}$ or nearly seventy-one per cent of the maximum value (amplitude) for a sine wave form.
period   the time taken for one complete cycle of the a.c.; i.e. so many seconds or milliseconds.

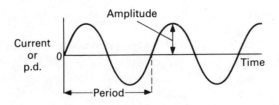

Fig. 17.1   Characteristics of an a.c.

frequency   the number of complete cycles per unit time, i.e. second, the reciprocal of the period, measured in *hertz* (named after the German physicist, Heinrich Rudolf Hertz, d. 1894) which means cycles per second; e.g. for the mains supply the period is 1/50s and the frequency 50 Hz.

phase   this idea is mentioned in section 13.7 in connection with a.c. generators; it is mainly used to describe how much out-of-step one a.c. is compared with another, i.e. *phase difference* between two alternating quantities is what matters (see section 17.3).

shape   the actual pattern of the time graph itself; usually the smoothly varying sinusoidal (or *sine wave*) shape is used, but in more complex cases square or triangular or pulsed shapes might be encountered.

17.2   Frequency and period

Either of these terms can be used to indicate the rate of the to-and-fro nature of an alternating current or voltage, in the same way that we talk about a bus service either as three per hour or one every twenty minutes (i.e. ⅓ hour).

In most countries the domestic supply uses a frequency of 50 or 60 Hz. Radar, radio and television, radioastronomy, microwave devices, etc., use a range of different a.c. frequencies from a few kHz to many MHz or even GHz (thousand million).

17.3   Phase and phase difference

The idea of *phase* in a.c. is best regarded as 'in-step-ness', and *phase difference* as the extent to which two alternating currents or voltages are out of step with one another. Figure 17.2 shows the time-graphs of pairs of alternating currents, both *in* phase and *out of* phase, but of the same frequencies.

Two graphs which are *in phase* keep step with each other exactly, reaching their maxima at the same times and crossing the zero line together. Being *out of phase* means not reaching maxima or not crossing the zero line at the same times. The special case of *anti-phase* occurs when the graphs move in opposite directions but still cross zero together.

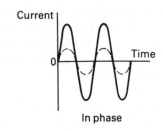

In phase

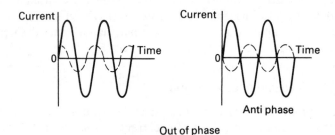

Out of phase

Anti phase

Fig. 17.2   Time graphs of alternating currents in and out of phase

The graphs can be out of step by intermediate amounts between completely in or out of phase, and the size of the time gap between them is called the *phase difference*. It is usually quoted as an angle in degrees between 0° and 360° which represents the span of one complete cycle (Figure 17.3).

Thus 90° relates to ¼ cycle, 180° to ½ cycle and 120° to ⅓ cycle, and so on. The phase difference between two time-graphs is the

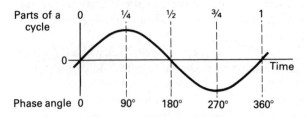

Fig. 17.3   Phase angles over a complete cycle

fraction of a cycle or the angle between corresponding points on the graphs, e.g. when they cross the zero line or when they reach their maximum values. Figure 17.4 shows some examples.

17.4 Lag and lead

For each of the graphs in Figure 17.4, the phase difference is indicated with reference to curve A and the angles shown go forwards from it. It is equally valid to use curve B as the reference because it is the difference that matters, but if that were done in the first example the angle on the graph would be 270° instead of 90°. To avoid such confusion it is necessary to say which of the curves is ahead of or behind the other, i.e. which *leads* and which *lags*.

Referring to the first example, curve A is 90° ahead of curve B since it reaches its maximum before B does, and in the second example it leads by 120°. In the third example it would be as easy to say that curve B is 90° ahead of A rather than 270° behind it. To be unambiguous the phase difference should be given as an angle together with a statement of which one is leading or lagging. (For a phase difference of 180°, lag and lead does not matter, of course, because the two would be in anti-phase.)

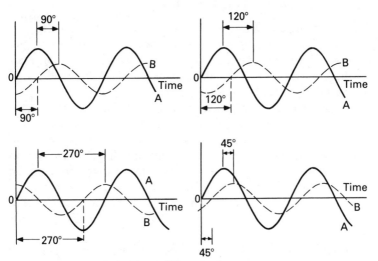

Fig. 17.4 Phase differences between two a.c.s

17.5   Phase with different frequencies

With two alternating currents of different frequencies the idea of phase difference is not so useful because, as Figure 17.5 shows, the two graphs are sometimes in phase and sometimes not. At times P and T, the two graphs are in phase, at R they are out of phase, and at between times they have different phase differences (e.g. at Q and S).

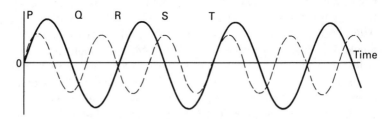

Fig. 17.5   Phase relationship with different frequencies

17.6   Size of a.c. – r.m.s. values

Using a moving-coil ammeter or voltmeter with a.c. is not effective because the movement to and fro is too rapid for the coil and needle to follow, and the result is that the average value of zero is indicated. Yet there must be a way of measuring and of describing the size of an alternating current or voltage.

Both a.c. and d.c. cause a wire to get hot. The heating effect is therefore used to indicate the size of an a.c., so that a current of '5 A' produces the same effect whether it is a.c. or d.c. We know from Chapter 8 that the square of the current (or voltage) is involved in the heat transfer, so for an a.c. it will be the average of its squared values that matters, i.e. its *mean square value*. We do not talk about an a.c. of 25 square amps, though, so it has to be the square root of the mean square value which is quoted, i.e. its *root mean square (or r.m.s.)* value.

Suppose we could take instantaneous readings for an a.c. over one cycle, we might get $0, 1, 2, 3, 2, 1, 0, -1, -2, -3, -2, -1$ amps.

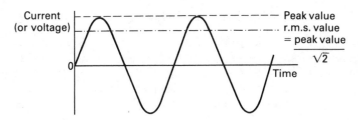

Fig. 17.6 Peak and r.m.s. values for a sinusoidal a.c.

The simple average of these readings is obviously zero, but by squaring them before averaging we could find the mean square value: 0, 1, 4, 9, 4, 1, 0, 1, 4, 9, 4, 1 total 38, mean square value $38/12 = 3.17$. The root mean square value would be $\sqrt{3.17} = 1.78$ A. (More than twelve readings are needed to find a reliable r.m.s. value, of course, but this illustrates the idea.) For this example, the a.c. supply would give the same heating effect as a d.c. supply of 1.78 A, so we call the a.c. also 1.78 A.

For a sinusoidal wave it can be shown mathematically that the r.m.s. value is $1/\sqrt{2}$ or 70.7 per cent of the peak value (Figure 17.6). Thus an alternating current quoted as 5 A (r.m.s.) would actually reach $5 \times \sqrt{2} = 7.07$ A in each direction, but would have the same heating effect as a direct current of 5 A. Similarly the 240 V mains varies from $+240 \times \sqrt{2}$, i.e. $+340$ V to -340 V.

17.7 Resistors and a.c.

If a.c. is applied to a circuit consisting of a resistor, the effect is that of Ohm's Law (Figure 17.7). Changes of voltage produce changes of current in phase with them, and the size of the current is always $I = V/R$ at each instant.

17.8 Capacitors and a.c. – reactance

With an alternating voltage applied to a capacitor (section 15.7) there is a phase difference of 90° between current and voltage with

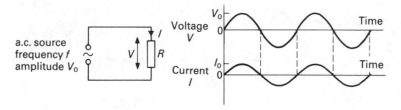

Fig. 17.7   a.c. circuit with resistance

the current leading (Figure 17.8). A detailed analysis shows that the amplitudes of the current and voltage are related by:

$$I_0 = \frac{V_0}{\dfrac{1}{2\pi f C}}$$

This indicates a *reactance* of $\dfrac{1}{2\pi f C}$ by the capacitor. This quantity

is a purely a.c. effect but it will be measured in ohms if f is in hertz and C in farads. So for a 1 μF capacitor at 50 Hz, its reactance would

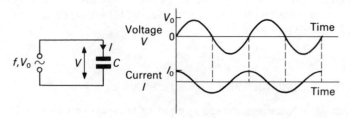

Fig. 17.8   a.c. circuit with capacitance

be $\dfrac{1}{2\pi\,50\,10^{-6}}\,\Omega$ = 3.2 kΩ. The symbol for capacitative reactance is X_C.

The interesting property of reactance is that it varies with the frequency of the a.c.; thus if f were 5000 Hz, the same 1 μF capacitor would have a reactance of only 3.2 Ω. This means that capacitors conduct high frequencies much more easily than low ones and can be used to filter out the latter from a mixture of frequencies.

17.9 Resistors and capacitors – impedance

If resistance and capacitance are both present, as in Figure 17.9, the general effect is similar but some details are altered. There is still a phase difference, but less than 90°, between current and voltage, and the relationship between their amplitudes is more complex. The current leads the voltage by an amount which depends on the product CR. The amplitude of the current is given by:

$$I_0 = \frac{V_0}{\sqrt{R^2 + \left(\dfrac{1}{2\pi f C}\right)^2}} \quad \text{or} \quad \frac{V_0}{\sqrt{R^2 + X_C{}^2}}$$

The whole of the denominator, $\sqrt{R^2 + X^2}$, is called the *impedance* of the circuit, symbol Z, also measured in ohms. The impedance (Z) is the combined resistive (R) and reactive (X) components of the circuit elements R and C.

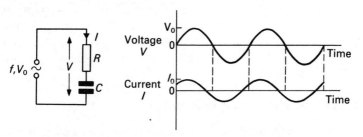

Fig. 17.9 a.c. circuit with resistance and capacitance

One curious point about this circuit is that the voltages across R and C will have a 90° phase difference. If an a.c. voltmeter (which shows r.m.s. values irrespective of phase) is connected in turn across R and C, the sum of its two readings will be greater than the applied voltage V!

17.10   Inductors and a.c. – reactance

From section 16.6 we know that the effect of an inductor is to cause the alternating current to lag behind the applied voltage by 90° (Figure 17.10). A full analysis in this case shows that the amplitude of the current is given by:

$$I_0 = \frac{V_0}{2\pi f L}$$

where $2\pi f L$ is the reactance X_L of the inductor (cf. reactance of a capacitor – section 17.8). A one-henry choke at 50 Hz has a reactance of $2\pi\, 50\, I = 314\ \Omega$ but, in contrast to the capacitor, the inductive reactance *increases* with frequency – at 5000 Hz the same choke would have a reactance of 31.4 kΩ. An inductor can therefore be used to filter out high frequencies from a mixture of frequencies.

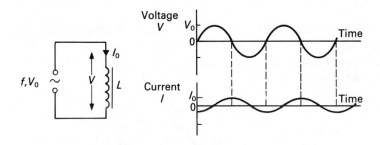

Fig. 17.10   a.c. circuit with inductance

17.11   Resistors and inductors

All chokes have some resistance of their own, so a more realistic case is that of a resistor and an inductor in series (Figure 17.11). This

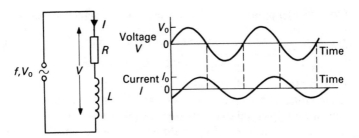

Fig. 17.11 a.c. circuit with resistance and inductance

time the phase lag is less than 90°, depending on the ratio of L/R, and the amplitude of the current is:

$$I_0 = \frac{V_0}{\sqrt{R^2 + (2\pi fL)^2}} \quad \text{or} \quad \frac{V_0}{\sqrt{R^2 + X_L{}^2}}$$

As with a capacitor, the term $R^2 + X_L{}^2$ is called the impedance of the circuit and is made up of a resistive and a reactive element. The peculiar trick of the total p.d. across R and L separately being apparently larger than the applied voltage happens here too.

17.12 Capacitors and inductors – resonance

The most important circuits in this chapter are those which include both capacitance and inductance (Figure 17.12). A resistor is shown, too, because the inductor will inevitably have its own resistance.

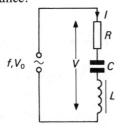

In this case the current might lag or lead the voltage depending on the sizes of L, C and R by anything up to 90°.

The amplitude of the current turns out to be given by:

Fig. 17.12 a.c. circuit with L, C and R

$$I_0 = \frac{V_0}{\sqrt{R^2 + \left(2\pi fL - \dfrac{1}{2\pi fC}\right)^2}} \quad \text{or} \quad \frac{V_0}{\sqrt{R^2 + X^2}}$$

The interesting point here is the special case which arises when $2\pi f L = \dfrac{1}{2\pi f C}$. At the frequency when this occurs the reactances of L and C cancel out, leaving the circuit in effect with only R controlling the current. Furthermore, under these conditions current and voltage are in phase with one another and the impedance Z is at its minimum equal to R. Figure 17.13 shows how I_0 varies if the frequency is made to rise from below to above the special value – called the *resonant* frequency.

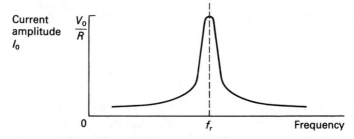

Fig. 17.13   Variation of current amplitude with frequency

The value of the resonant frequency f_r is calculated from:

$$2\pi f_r L = \frac{1}{2\pi f_r C}$$

which gives

$$f_r = \frac{1}{2\pi \sqrt{LC}}.$$

For $L = 1$ H and $C = 1\,\mu\text{F}$, f_r works out at almost 160 Hz, whilst for 10 mH and 100 pF the value is 0.16 MHz.

Examples of resonance can be found in mechanical and acoustical systems of various kinds:

the rattling of part of the bodywork of a car at a certain engine speed;

the much louder response of a poor quality loudspeaker to certain musical notes;

the large amplitude of a child's swing if pushed at just the right moments;

the true story of a wine glass being shattered by an operatic
singer reaching a high note;

the catastrophic collapse of a suspension bridge caused by
vibrations at a certain wind speed;

In each of these cases a large amplitude of vibration occurs only at a
certain frequency, and it is one at which the system would vibrate
naturally if given a slight disturbance. The natural oscillations build
up as they are reinforced by continual disturbances at just the right
times. The peak amplitude of Figure 17.13 becomes very much
larger at the resonant frequency than at any other, and in the case of
the capacitor-inductor circuit the sharpness of the peak depends on
the value of the resistor – the lower the R the sharper and higher is
the peak.

17.13   Capacitor-inductor circuit

If a capacitor is charged up and then discharged through an inductor
(Figure 17.14) the current flows first one way and then the other,
backwards and forwards at a particular frequency which turns out to
be the same as the resonant frequency described in section 17.12.

If there were no resistance, these oscillating currents would carry
on for a long time, but heat is developed in a resistor which
gradually reduces the total energy of the system. This is very like
a swinging pendulum which gradually slows down because of fric-
tional effects, but which, if given a small jolt at the right point in

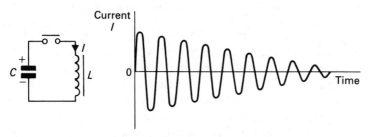

Fig. 17.14   Oscillating current in an *L-C* circuit

its swing can be kept going at a steady amplitude and at its natural frequency. An *L-C* parallel circuit can be made to behave like this too by using the small signals picked up from a radio or TV aerial (Figure 17.15).

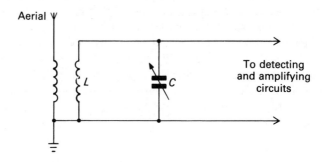

Fig. 17.15   Use of an *L-C* circuit in a radio aerial

Because of its own natural frequency, a given *L-C* combination will select that particular frequency for a high response rather than any others picked up by the aerial, so providing a means of tuning in to an individual broadcasting station. A variable capacitor then allows a range of stations to be chosen at will.

17.14   Oscillators

An important use of the *L-C* circuit is to generate electrical oscillations over a range of frequencies. *Oscillators* are electronic devices which are designed to do this, and their basis is the resonating property of an *L-C* coupling. What is needed is a way of starting an *L-C* circuit oscillating and then feeding sufficient energy into it at its own natural frequency to maintain the oscillations. Without going into the details of transistor circuitry, Figure 17.16 shows one arrangement which would achieve this result.

Oscillations are started in $L\text{-}C_1$ by a contact with the $+9$ V supply, but they are subsequently maintained by the action of the transistor. By varying C_1 the frequency of oscillation can be altered. A good

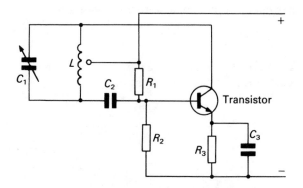

Fig. 17.16 Circuit of a Hartley oscillator

sinusoidal wave form can be obtained from circuits with *L-C* components.

One obvious application of electronic oscillators is in the generation of sounds for electronic organs and synthesizers.

17.15 Summary

Characteristics of an a.c. are amplitude, r.m.s. value, period, frequency, phase and shape.

Phase differences are described as leading or lagging by fractions of a cycle or of 360°.

The r.m.s. value of an a.c. is the d.c. equivalent which produces the same heating effect.

For a sine wave, the r.m.s. value is 70.7% of the peak value.

a.c. through a resistor suffers no change of phase.

a.c. through a capacitor suffers a 90° phase change, current leading.

The reactance of a capacitor $X_C = \dfrac{1}{2\pi f C}$

The impedance of a circuit $Z = \sqrt{R^2 + X^2}$

a.c. through an inductor suffers a 90° phase change, current lagging.

The reactance of an inductor $X_L = 2\pi f L$.

With resistance included, phase changes are less than 90°.

a.c. through an *L-C* circuit leads to resonance at a
frequency of $f_r = \dfrac{1}{2\pi \sqrt{LC}}$

L-C circuits are used for selecting particular frequencies.
L-C circuits form the basis of oscillators.

18

Communicating with Electricity

18.1 Elements of communication

All systems of electrical communication over large distances have common features which in practice might be quite simple or very complex (Figure 18.1).

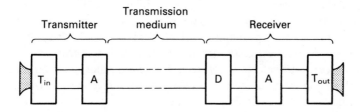

T_{in} input transducer which converts information to be transmitted into electrical signals;
A   amplifiers to increase the strength of small signals if possible without distortion or loss;
D   detector/decoder which picks up the transmitted signals, and isolates the essential parts of them;
T_{out} output transducer which converts the electrical signals into an appropriate form for use.

Fig. 18.1   Components of a communications system

This model is an over-simplification but does indicate the main parts of any communications device. A very simple example not using electricity is the toy telephone made from two metal cans and

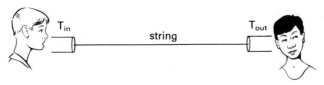

Fig. 18.2   A string telephone

a piece of string (Figure 18.2). Here one metal can converts the sound into mechanical vibrations which travel along the string (which must be kept taut) and the other can converts them back into sound. The cans also offer a certain amount of amplification and, provided the string does not touch anything on the way, the 'telephone' works surprisingly well.

18.2   Some practical applications

The field of telecommunications is a highly developed and very sophisticated one and largely beyond the scope of this book, but a few examples and principles can be dealt with. The telephone system is in essence a simple arrangement: a microphone is the input transducer, a transformer acts as a simple amplifier, wires are the transmission medium and a loudspeaker is the receiver and output transducer. The switching and connecting circuits are really a convenience to widen the scope of the service, and not part of the basic system. Because there is physical connection through wires from transmitter to receiver in most of the telephone network there are limits to how many links can be accommodated at one time.

Radio and TV introduce more complexities because the medium used is not a solid one but electromagnetic waves which travel across large distances (even astronomical ones). Sound and visual information have to be converted into electrical signals which are carried by other signals from the transmitting stations. The receiver then needs an aerial to pick up the signals and very complicated electronic devices to decode and present the original information with suitable amplification in the process.

Teleprinter services are similar to radio but the information is transduced from written material in coded form and transmitted over fairly short distances by line-of-sight short wave signals, with booster stations which relay them from one place to another. Special decoding machinery is needed which results in the output being printed automatically.

Morse code is one of the simplest ways of coding information and puts it in a particularly easy form for transmission and decoding. Digital techniques with electronic scanners are the modern development from that original idea of separate pulses to represent pieces of data.

18.3 Microphones

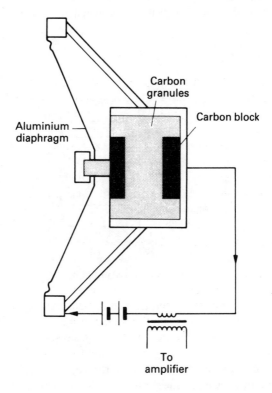

Fig. 18.3 A carbon microphone

Carbon granule   The resistance of loosely packed carbon granules in a sealed container varies if different pressures, such as those caused by a sound wave, are applied. Figure 18.3 shows such a device through which a current is passed. Changes in the current are caused by changes of resistance and the transformer passes them on (but not the steady current) for transmission.

This type of microphone needs a d.c. supply from which the power comes to produce the a.c. component which is the transduced signal. Early telephone handsets incorporated such microphones, the d.c. supply coming from batteries in the local exchange.

Moving-coil   A different type of microphone uses a small coil of wire attached directly to the diaphragm, which oscillates under the influence of the sound vibrations between the poles of a magnet (Figure 18.4).

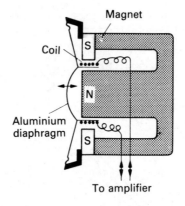

Fig. 18.4   The moving-coil microphone

No electricity supply is needed for this microphone because induced e.m.f.s are produced by the coil's movements which can be fed straight to an amplifier.

Condenser Another design uses the diaphragm itself as one plate of a capacitor (previously called a condenser) whose capacitance is varied by slight movements of the diaphragm. In a d.c. connection (Figure 18.5) charge movements occur which cause small voltage changes that can then be amplified.

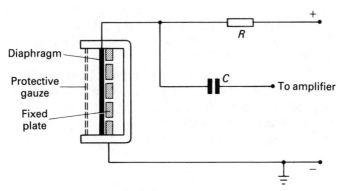

Fig. 18.5 A condenser microphone

The capacitor C blocks the d.c. component of the p.d. across the microphone and conducts only the changes of voltage. This type of microphone is often incorporated into portable tape recorders.

Ribbon A corrugated aluminium ribbon is suspended edgeways between the poles of a magnet (Figure 18.6) and when it vibrates under the influence of a sound an e.m.f. is induced which can be amplified and transmitted.

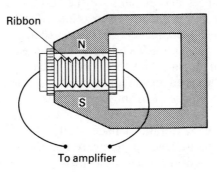

Fig. 18.6 A ribbon microphone

18.4   Loudspeakers

Microphone design has led to a variety of forms which suit a range of applications, and excellent fidelity can be obtained from the best of them, giving faithful reproduction in electrical signals of the sound inputs. Loudspeakers, on the other hand, have been dominated by one type, the moving-coil loudspeaker described in section 9.9, which is virtually the same device as the moving-coil microphone with a large flexible cone in place of the diaphragm (Figure 9.16, cf. Figure 18.4).

For the highest quality of sound reproduction several loud-speakers would be built into a single unit to deal with the full range of frequencies in sound, from 20 Hz to 20 kHz. Listening to a full symphony orchestra through the small single speaker of a television set gives a poor impression of the real sound.

18.5   Transmission through wires

Apart from Morse code signals, which are just pulses of current, electrical signals consist of alternating currents or voltages. When a current is switched on its effects appear to be instantaneous, yet however fast it seems, the greatest possible speed would be that of light, i.e. 3×10^8 metres per second (or 186 000 miles per second). This means that in the case of alternating currents being transmitted over long distances it is possible for the current to be reversing direction at one end of a cable before its first effect had reached the other end. Suppose a frequency of 10 kHz was being transmitted – in $\frac{1}{10000}$ second it goes through a full cycle of changes backwards and forwards, but in the same time the signal would progress only thirty kilometers at the speed of light, about nineteen miles.

For high frequencies and long cable runs the signals will be travelling along the wires as a series of waves, just as waves can be sent along a rope by vibrating one end of it by hand. Transmission cables have to be designed to minimise the effects of resonances (section 17.12) because even single wires have capacitance and inductance. Another problem is the tendency for high frequency a.c. to travel along the outsides of conductors only. (More advanced textbooks should be consulted for further consideration of this topic.)

18.6 Transmission without wires

The possibility of wire-less transmission was realised in 1865 by British physicist James Clerk Maxwell (d. 1879) when the link between electric and magnetic fields was established (that each can be generated by changes in the other irrespective of the presence of material substances – see Chapters 7 and 10). The actual demonstration of the transmission was first made by Heinrich Hertz in 1888 using a spark transmitter and receiver (Figure 18.7) over the length of a laboratory.

The induction coil (section 10.4) produces an intermittent pulse of electricity each time its primary circuit is interrupted by the contact breaker, which gradually charges up the large metal plates A and B. When their potential difference reaches a high enough level a spark passes across the gap between X and Y. There was great excitement when the much smaller sparks were seen at the gap in the loop of wire C which acted as the receiver. Hertz had tried

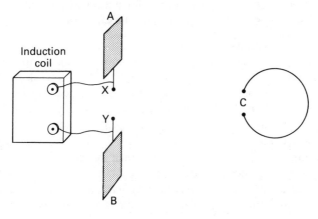

Fig. 18.7 Arrangement for the first radio transmission

several arrangements before finding the one which produced the effect he felt sure was possible, namely the use of *electromagnetic waves* (radio waves) to carry energy between transmitter and receiver without the need for wires.

In Hertz's experiment the radio waves contained a mixture of many frequencies just as a sudden noise consists of a mixture of sound frequencies, but it is possible now to generate any desired frequency of radio waves for transmitting signals over large distances. Appropriate aerials for sending and receiving are needed. These vary in size according to the frequency, low frequencies requiring longer aerials than high frequencies.

18.7   Modulation of radio waves

The purpose of radio transmission, for example, is to enable sounds to be conveyed between one place and another. Sound waves, if merely converted into radio waves of the same frequency for transmitting across the country, would require vast aerials hundreds of metres or more in length. For this reason a radio wave of much higher frequency is used to carry the sound signal. The radio receiver is designed to pick up the radio wave, separate the frequencies and amplify only the sound frequencies. The radio wave is *modulated* with the sound wave at the transmitter to enable the process to work. Another advantage of doing this is that the radio receiver can select one particular radio station's frequency rather than picking up everything in the air at the same time.

Amplitude modulation (AM)   In this kind of modulation, the amplitude of the radio wave (called the *carrier*) is made to vary at the frequency of the sound. Figure 18.8 shows the principle of what is meant by this, but the scale is hopelessly wrong because the radio frequency might be 1 MHz whilst the audio frequency is only 1 kHz, i.e. there should be a thousand oscillations of the radio wave to one of the sound wave.

The task of the radio receiver is to pick up the radio frequency wave but to respond then to the changes in its amplitude to obtain the sound information it carries. *Amplitude modulation* (AM) is used with the long, medium and short wavebands of ordinary radio sets.

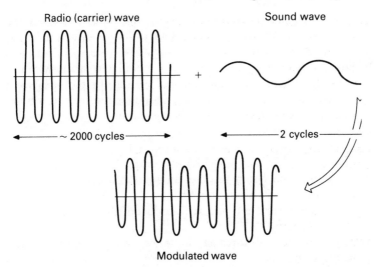

Fig. 18.8 Amplitude modulation

Frequency modulation (FM)   An alternative way of mixing up the sound information with the radio frequency is to modulate the frequency of the latter rather than its amplitude. Figure 18.9 attempts to show this effect, again not to scale.

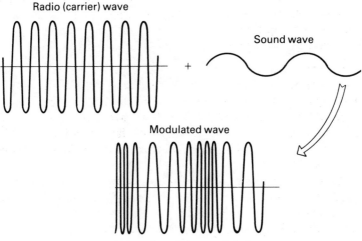

Fig. 18.9 Frequency modulation

With frequency modulation the receiver needs to be able to respond to the frequency changes in order to disentangle the sound information from the radio wave. The VHF waveband and television transmissions use this type of modulated carrier wave.

For further details of detection and decoding of radio signals and of associated electronic circuits the reader is referred to a book dealing specifically with AM and FM reception.

18.8   Summary

A communication system requires an input transducer, a transmitting medium and an output transducer.

Microphones of various types are transducers for sound energy into electrical energy.

Transmission through wires has limitations at high frequencies and over large distances.

Wireless transmission involves electromagnetic waves.

Sound frequencies are transmitted by amplitude or frequency modulation of radio frequency waves.

19

Basic Electronics

19.1 A growth industry

The development of modern electronics can be traced back to the discovery of 'semiconducting' materials and the invention of the transistor in the late 1940s. Since then the ability to produce large quantities of the substances which make up these components, and the skill in designing smaller and smaller circuitry in which to use them, has resulted in a real electronics revolution. The rate of progress since 1950 has been accelerating steadily so that it is possible for new ideas to be conceived, developed, tested and put into production in such a short time that few people are able to keep up to date except in very specialised areas.

It is easy to take electronic devices for granted now they have become such common features of industry, entertainment, commerce, education and leisure. Items such as transistor radios, electronic games, home computers, video recorders, automatic cash registers, robot-operated manufacturing processes, electronic timers, digital clocks, calculators, electronic organs, synthesizers, lasers, infra-red cameras, cardiac pacemakers . . . this list repesents just part of the range of appliances in which electronics is put to practical use. This chapter will be able only to introduce the reader to the basic elements of many electronically operated devices.

19.2 Semiconductors: n-type and p-type

Most metals, especially copper and silver, conduct electricity very easily whilst the plastic materials like alkathene, perspex and PVC are used as insulators because they hardly conduct at all. Substances which behave electrically in an intermediate way between conductors and insulators are called *semiconductors*: their resistances are usually fairly high and variable. The elements germanium and silicon are examples of semiconducting materials whose resistivities are high but fall markedly if they are heated.

A more useful material can be obtained by adding very small quantities of other elements such as phosphorus, arsenic or boron to germanium or silicon, which greatly reduce their resistances. Two different types of semiconductors, called *n-type* and *p-type*, can be made this way: for example, germanium with phosphorus as an impurity is n-type, and silicon with arsenic is p-type. (The names n and p refer to the **n**egative and **p**ositive charge carriers which conduct the electricity, negative carriers predominating in n-type and positive in p-type.) Figure 19.1 illustrates the basic properties of the two types of semiconductor.

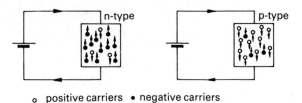

o positive carriers   • negative carriers

Fig. 19.1   n- and p-type materials

19.3   The pn junction

Many semiconductor devices depend on what happens when n-type and p-type materials are joined together. Without going into the mechanism too closely, the main effect is that conduction is very much easier in one direction than in the other (Figure 19.2). One use of such an effect is a one-way device like a diode (Chapter 12).

19.4   The junction diode

A pn junction between pieces of p- and n-type materials together with two connecting wires and encased in a suitable container forms a diode. The connection marked + is called the *anode* and that marked − the *cathode*. The conduction curves for a pn junction

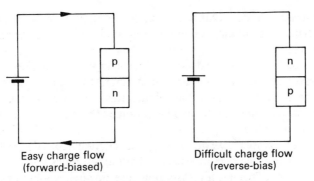

Fig. 19.2 Current flow in a pn junction

diode (Figures 4.3 and 12.2) are shown in Figure 19.3 (note the different scales on the + and − axes).

Such a diode would be described by its average forward current (e.g. 1 A) and its maximum reverse voltage (e.g. 50 V), neither of which should be exceeded under normal conditions. Silicon diodes are usually preferred for rectifiers because of their higher values for each of these quantities.

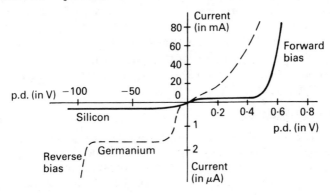

Fig. 19.3 Conduction curves for a pn diode

19.5 Other diodes

Reference to books on electronics will show other kinds of diodes designed for specific uses, some of which are listed and summarised here:

Point contact diode   This is used to detect radio signals, where it acts as a combined capacitor and diode, usually germanium based.

Zener diode   This is used as a stabiliser in power supplies, used under reverse voltage at a point where the current does not effect the p.d. across the diode; this reference voltage (or Zener voltage) can be fixed in a range of values between, say, 3 V and 200 V.

Diac   This consists of two Zener diodes connected back to back which conduct in either direction at the reference voltage.

Photo diode   This has a transparent case through which light can enter; the brightness of the light (and its colour) affects the reverse current; it is used in fast counters or punched tape readers and in light meters.

Light emitting diode (LED)   This is made of a gallium-arsenic-phosphorus compound which emits light when conducting under forward voltage; used as indicator lamps in digital circuits, e.g. calculators or display units.

Varicap diode (varactor)   This is a combined diode and capacitor with a range of capacitance depending on the reverse voltage.

Gunn diode   This is made of n-type gallium arsenide, and is used in microwave circuits.

Photo voltaic cell   This is a pn junction which produces an e.m.f. when light falls on it; used in panels for artificial satellites and to support small devices such as electronic watches or calculators.

19.6   Transistors

The basis of modern technology was the development of the *transistor*, which is a three-element device, either a pnp combination or an npn. The principal applications of transistors centre on their abilities to act as amplifiers and switches, i.e. active circuit components compared to passive ones like lamps, resistors or diodes. They are immediately suitable for control circuits, logic systems and radio equipment, especially since it is possible for thousands of transistors to be mounted with other circuit elements on a small 'chip' of silicon about 5 mm square (section 19.17).

Single transistors come in various sizes and shapes with three connecting wires or points (Figure 19.4).

Fig. 19.4   Types of transistor

There are two basic types of transistor, the (ordinary) junction transistor and the field effect transistor (FET).

19.7   Junction transistors

Figure 19.5 shows the two possible arrangements, pnp and npn, and their circuit symbols. In each case a *base* of one type of semiconducting material is sandwiched between two pieces of the other type called the *collector* and *emitter*. The npn transistor is usually made of silicon and the pnp of germanium. The arrow indicates the direction of (conventional) current between base and emitter.

Note: The construction of transistors is much more complex than a simple double junction, the action depending upon the amount of impurity elements in each part and the thickness of the base.

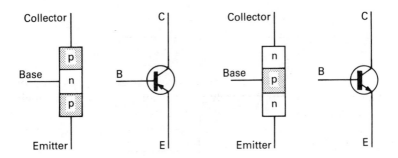

Fig. 19.5   Junction transistors

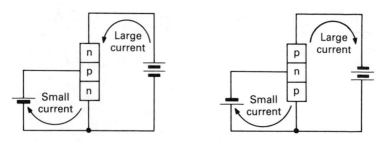

Fig. 19.6 The basic transistor action

19.8 The transistor action

For the transistor to operate, the two junctions must be operated under different voltage conditions – the base-emitter junction forward and the collector-base reverse, as in Figure 19.6. (Although a cell is shown providing the necessary p.d. between base and emitter, in practice the voltage would be produced in other ways.)

The action itself is that the small current between base and emitter can control a much larger current between collector and base, including switching it on and off. The sizes of typical currents are indicated on Figure 19.7. (Note the resistor R, say 1 kΩ, which limits the current in the base.)

If the base current (0.05 mA) were to be interrupted, there would be no current into the collector.

The junction transistor is a current-operated device in which one

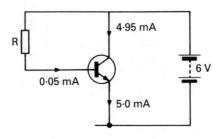

Fig. 19.7 Typical current sizes

current can switch on and off, or alter the value of, a much larger current, often between ten and one thousand times its size, depending on the particular transistor. Important quantities are the size of each current, the ratio of them and, for safety's sake, the maximum base current.

19.9   Field Effect Transistors (FET)

In this type of transistor, which also has three terminals, it is the voltage of the *gate* (G) which controls the current flowing between the *drain* (D) and the *source* (S). The different behaviour of the FET is due to the way the n- and p-type semiconducting sections are constructed.

Figure 19.8 shows the symbol for an FET and typical sizes of currents and voltages. The important quantity is the change produced in the drain current for a given change in the gate voltage.

The way an FET works means that it is a voltage-operated device, which makes it very useful for handling the output of a crystal pick-up of a record player, for example. A particular virtue of certain FETs is that they can very easily be manufactured in large numbers on a small silicon chip to form a single unit called an *integrated circuit* or microprocessor (section 19.17).

Two different types of FET are in common use, the junction-gate FET (or JUGFET) and the metal oxide semiconductor FET (or MOSFET).

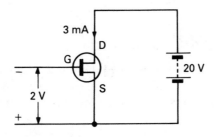

Fig. 19.8   Typical FET quantities

19.10   Transistor switching

The principle of the transistor's use as a switch can be seen from the circuit of Figure 19.9.

The 1 kΩ and 10 kΩ resistors act as a potential divider, placing the base of the transistor at $^{10}/_{11}$ of 6 V in (*a*) and $^{1}/_{11}$ of 6 V in (*b*). The first case gives a current into the base which switches the transistor 'on', and allows the current to pass through the collector and emitter (as in Figure 19.7). In (*b*) only a tiny current would pass into the base, not enough to allow the collector to conduct, and the transistor would be 'off'. (The resistor R is included to prevent too large a base current from damaging the transistor.) To investigate the exact point when the transistor changes from 'on' to 'off', the circuit of Figure 19.9(*c*) could be used, the voltmeter V indicating the potential of the base as the variable resistor is changed from 10 kΩ to zero.

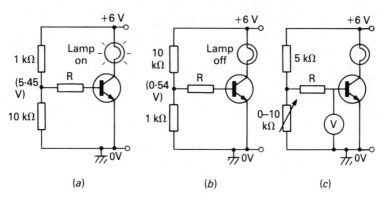

Fig. 19.9   The transistor as a switch

19.11   Light-operated switches

The two circuits of Figure 19.10 show how a simple light-operated alarm circuit can be made using a light dependent resistor (LDR) which has a high resistance when dark but a much lower resistance in bright light.

The lamp would light in the first circuit in dark conditions but in the second circuit in light conditions (compare with the circuit of

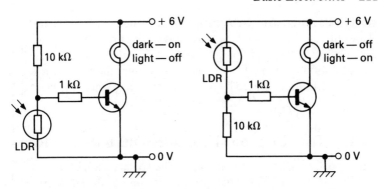

Fig. 19.10 Circuits for light-operated alarms

Figure 19.9). A warning would be given by the lamp if, say, an intruder switched room lights on or if a lighting system were obstructed. The warning lamp could be placed some distance from the LDR if necessary, and infra red light beams could be used.

The same circuit could be used to detect fire or frost if the LDR were replaced by a thermistor (a semiconducting material whose resistance varies greatly with temperature – section 19.18), and the lamp could in turn be replaced by a buzzer or a light emitting diode (LED). In the same way a relay could be used in place of the lamp (Figure 19.11). Relays are electromagnetic switches in which a small

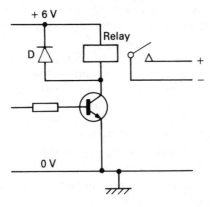

Fig. 19.11 Use of a relay with a transistor switching circuit

current activates an electromagnet which attracts a piece of iron that is part of a switch in a separate circuit such as a fire alarm. The alarm could be some distance away from the rest of the circuit. (In such a circuit a diode D would be required to protect the transistor from the back e.m.f. generated by the coil in the relay when it is switched off by the transistor – compare section 10.12.)

19.12   Transistors and logic gates – the binary system

An important use for transistors is in the construction of integrated circuits (section 19.17), which in turn are one of the basic components of computers and microprocessors. Devices like these are really very fast electronic calculating machines which can handle millions of small mathematical steps in a much shorter time than any human could manage. However, although we use the decimal system to count in tens, hundreds and thousands, computers work in the *binary* system, which is based on two digits.

The decimal system requires ten digits (0, 1, 2, 3, 4, 5, 6, 7, 8, and 9) from which all numbers can be composed. The position occupied by a digit in a number is just as important as the digit itself, so that 43 implies 4 tens and 3 ones and 618 represents 6 hundreds, 1 ten and 8 ones. When fractions are written the same system applies, with 0.27 being 0 ones, 2 tenths and 7 hundredths. (The early teaching of mathematics to young children emphasises the importance of place-value as well as the sizes of the digits themselves.)

In the binary system there are only two digits, 0 and 1. Place-value is just as important, however, but the positions are worth one, two, four, eight, sixteen, etc., instead of one, ten, hundred, thousand. On this system, then, the number 1011 means 1 eight, 0 four, 1 two and 1 one, i.e. eleven on the decimal system. The reason why transistors can be used to count in binary numbers is that they have essentially only two states – being 'on' or 'off', having a 'low' output or a 'high' one. If these two states are made to correspond to 1 and 0 the transistor becomes a binary device. Circuits can be built in which all the normal mathematical processes can be performed, but in the binary rather than the decimal system. (It is necessary also, of course, to have ways of translating between decimal and binary numbers.)

A *logic gate* is a switching circuit which gives high or low outputs depending on the signals fed into it and can be used as the basic binary circuit. Several different types of gate can be made using the switching properties of transistors, some of which are described below.

NOT gate   its output is high when its input is *not* high; or low when its input is *not* low.

In Figure 19.12, if the input is high (say 5 V), the transistor will be switched 'on' and a large collector current will flow through the resistor R_L giving a p.d. across it which makes the output voltage low (i.e. very near 0V). If the input is low (say 0.5 V), hardly any current passes through the collector since the transistor is switched 'off', giving a small p.d. across R_L; the output voltage is therefore not much less than 6 V, i.e. high.

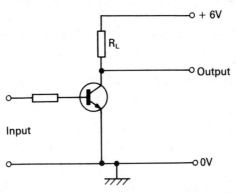

Fig. 19.12   A NOT gate circuit

The action can be summarised by a truth table in which 1 indicates high and 0 low, and the whole represented by a single circuit symbol (Figure 19.13).

Input	Output
0	1
1	0

Fig. 19.13   NOT gate: truth table and symbol

NOR gate   this gate has two (or more) inputs and its output is high only if neither one input *nor* the other is high. The circuit, the truth table and the symbol are shown in Figure 19.14. (From Figure 19.12 it is clear that a NOT gate is a single input NOR gate.)

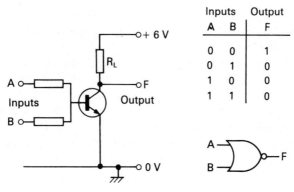

Inputs		Output
A	B	F
0	0	1
0	1	0
1	0	0
1	1	0

Fig. 19.14   NOR gate: circuit, truth table and symbol

OR gate   a gate whose output is high if either one input *or* another is high (or both are high). The circuit is a NOR gate followed by a NOT gate (Figure 19.15).

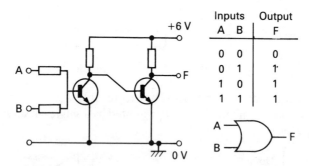

Inputs		Output
A	B	F
0	0	0
0	1	1·
1	0	1
1	1	1

Fig. 19.15   OR gate: circuit, truth table and symbol

NAND gate   a gate whose output F is *not* high when both inputs A *and* B are high (Figure 19.16).

AND gate   its output F is high only when both inputs A *and* B are high (Figure 19.17). (This is a NAND gate followed by a NOT gate.)

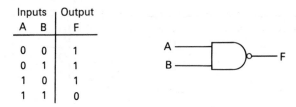

Fig. 19.16 NAND gate: truth table and symbol

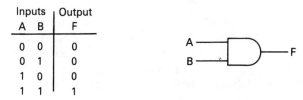

Fig. 19.17 AND gate: truth table and symbol

Exclusive OR gate a gate whose output F is high only if either input is high but not both, i.e. only if the inputs are different (Figure 19.18).

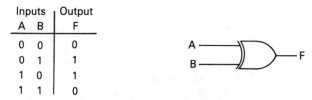

Fig. 19.18 Exclusive OR gate: truth table and symbol

Circuits have been included only for the simplest gates, and it is not always a matter of just adding on component circuits to make a more complex one, but the truth table approach is useful in deciding what a sequence of gates would actually do in practice. For example, if the set of NAND gates in Figure 19.19 is tackled by a series of truth tables, it will be found to be an exclusive OR gate.

In practice, all the separate NAND gates could be built into a single integrated circuit (section 19.17) with the facility to interconnect them in a variety of ways.

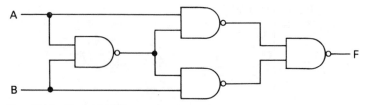

Fig. 19.19 Four NAND gates connected to make an exclusive OR gate

19.13   Uses of logic gates

The simple ideas described above soon lead to quite complicated circuits beyond the scope of this book. Suffice to say that combinations of logic gates are used in many digital electronic devices such as computers, decimal converters, decision-making circuits, encoders and decoders, adders, multiplexers, memory circuits and many more.

19.14   Transistor amplifiers

In section 19.8 the action of transistors in operation was seen to lead to two main uses: as switches and as amplifiers. The latter use was based on the control of a large collector current by a much smaller base current. Amplifiers can be designed in electronics to produce as output an enlarged version of an input, in terms of current, voltage or power and at audio frequencies (20 Hz to 20 kHz) or at radio frequencies (20 kHz to 10 000 MHz or more).

Transistors are used in most electronic devices at some stage or other. Whilst they require only small power sources themselves, the output of the loudspeaker or television tube can be very much greater. Signals of various kinds need to be amplified to be of much practical use: most obviously the radio or TV signals picked up by an aerial often need to be amplified before being decoded and converted into sound and vision with even more amplification.

The general symbol for an amplifier is:

Input   Output

often abbreviated to just   with the reference or zero level left unwritten.

19.15   Simple transistor amplifier (audio frequency)

The circuit of Figure 19.20 is the simplest example of a voltage amplifier.

If a small alternating voltage is applied as the input, a larger but inverted alternating voltage of the same frequency appears at the output. A cathode ray oscilloscope (section 6.8) connected to the input or output terminals would display the curves shown in Figure 19.20. An amplification of about seventy can be obtained in this way. Note the resistors R_B and R_L similar to those in Figures 19.7 and 19.12. The capacitor C serves to make the output voltage symmetrical about the zero level.

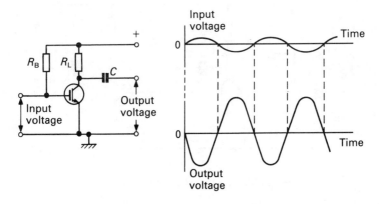

Fig. 19.20   Voltage amplification by transistor

19.16   Stabilised and two-stage amplifiers

Circuit diagrams for a simple temperature-stabilised amplifier and a two-stage voltage amplifier are shown in Figures 19.21 and 19.22. An amplification factor of several thousand could be obtained with the latter circuit. In practice the capacitors in both these circuits would be electrolytic ones. The output from the two-stage amplifier would not be inverted as in the single-stage one.

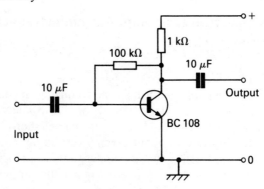

Fig. 19.21   A temperature-stabilised transistor amplifier

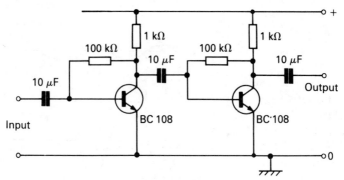

Fig. 19.22   A two-stage amplifier circuit

19.17   Integrated circuits (IC)

The techniques of manufacturing components for electronic circuits now allow for thousands of circuit elements – resistors, diodes, transistors, capacitors – to be mounted, complete with interconnections, on a silicon 'chip' about 5 mm square and 0.5 mm thick. The silicon has to be made to a purity of about one part in ten thousand million and in a cylindrical bar about 10 cm across. Wafers are cut from such a bar and a couple of hundred identical circuits are formed by a repetitive process alongside one another on each ½ mm-thick wafer. After individual testing of each circuit, the thirty per cent or so of successful ones are mounted in plastic cases with gold connecting wires added to the external pins. The whole IC would be about 2 cm long (Figure 19.23). The number of external connecting pins varies between eight and twenty-four, depending on the particular circuits contained on the chip.

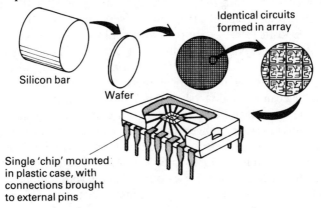

Fig. 19.23 Stages in the manufacture of a 'chip'

The advantages of an integrated circuit are its size, relative cheapness and reliability. Many different designs of IC are available and practical electronics has become much more a matter of using ready-made circuitry than laboriously building up circuits from separate components. Increasingly, too, the emphasis has shifted to the uses they can be put to rather than, for many people, the way they work. A 'systems approach' is the way electronics is tackled through ICs and pre-manufactured circuitry instead of through detailed knowledge of the characteristics of individual elements.

ICs form the basis of the development in computer design which has made it possible for both costs and sizes to come down to manageable proportions for commercial, educational and domestic use. There seems to be no limit to the ingenuity of the microelectronics industry to put on to a silicon chip almost any required circuitry. The vastly increased power of the large computers is another evidence of the advantages of miniaturisation of the basic components. No doubt yet newer techniques will in turn make these developments seem old fashioned, just as the radio valve has been made obsolete by the transistor, and the transistor by the integrated circuit.

19.18   Other semiconductor devices

Thermistor   Whilst metallic conductors generally increase their resistance when they are heated (section 4.13), the effect is not all that large – a doubling for a rise of, say, 300 degrees C. In the case of semiconductors (e.g. silicon) there is a very much larger effect, but in the opposite sense – their resistances fall when they become hot, and often by very large amounts. The thermistor Th3, for instance, has a resistance of 380 Ω at 25°C but only about 30 Ω when hot. Their main use is in electrical thermometers were changes in resistance can be made to show temperature changes directly, one practical advantage being that they can be made quite small and placed in awkward or dangerous positions well away from the rest of the circuitry and the operator. Their small sizes give the chance for very localised temperatures to be measured. Thermistors can also be made which give large rises of resistances with temperature, a property useful for protection of circuits against overheating.

Thyristor   This is a four layer pnpn sandwich with three contacts – *anode*, *cathode* and *gate*. Under forward voltage it conducts when the gate is positive and stays conducting on removal of the positive voltage until the anode voltage is removed or reversed. In the circuit of Figure 19.24 the lamp will light when both S_1 and S_2 are closed but then remains on when S_2 is opened.

In a.c. circuits a thyristor can be triggered at any positive voltage. The effect in that the thyristor conducts only for part of the a.c. cycle, as in Figure 19.25, the timing of the gate pulses deciding the point at which conduction begins.

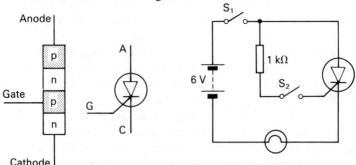

Fig. 19.24   Thyristor: symbol and circuit

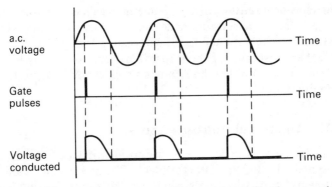

Fig. 19.25 Thyristor conducts part of the a.c. triggered by the gate voltage

By using a double 'back to back' thyristor, called a *triac*, conduction can be allowed on part of the negative half-cycle as well (compare half- and full-wave rectification – section 12.4). This type of current control is far more efficient than using a variable resistor since there is almost no power loss when the thyristor is off. Theatre lighting, for example, can be controlled using thyristors without the generation of large amounts of heat in the dimmers. The basic circuit for doing this is shown in Figure 19.26, in which the variable resistor controls the proportion of the a.c. cycle during which conduction occurs.

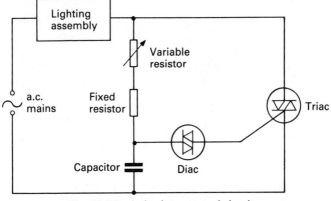

Fig. 19.26 A thyristor control circuit

Light dependent resistor (LDR)   Sometimes called a photoconductive cell, this is a resistor made from cadium sulphide whose resistance falls when illuminated externally. The ORP12, for instance, has a resistance of about 10 mΩ in darkness but only 1 kΩ in daylight – a factor of ten thousand. An application is illustrated in section 19.11.

19.19   Diode and transistor codes

Several systems of codes are in use to identify the hundreds of different diodes and transistors available. The most common are the Continental and American systems. In the former, a first letter indicates the semiconductor material used (A = germanium, B = silicon), one or two further letters show the appropriate use for which the device is designed (A = signal diode; Y = rectifier diode; Z = Zener diode; C = audio frequency amplifier; F = radio frequency amplifier; S = switching transistor), and a number relates to the electrical characteristics. BY118, for instance, is a silicon rectifier diode. Variations also exist within the system, e.g. an OA91 is a germanium point contact diode.

In the later American system, all diodes are coded 1N and all transistors 2N, followed by a four-digit number which refers to the properties of the devices. Examples are 1N4001 and 2N3053.

There are also letter-and-number codes for the various types of integrated circuits such as gate units, operational amplifiers, counters, etc.

19.20 Summary

Semiconducting materials can be n-type or p-type.

A pn junction conducts like a diode.

Different types of semiconductor diodes offer a variety of uses.

A transistor consists of a pnp or npn combination.

Junction transistors act as current-operated switches or amplifiers.

Field effect transitors (FETs) are voltage-operated devices.

LDRs used with transistors provide light-sensitive circuits.

Logic circuits comprise NOT, NOR, OR, NAND, AND and exclusive OR gates.

Transistor amplifiers need stabilising against temperature changes.

Integrated circuits (ICs) contain many elements mounted on a thin chip of silicon.

Other devices include thermistors, thyristors, Zener diodes, diacs and triacs.

Transistor codes indicate their construction and applications.

20

Conduction through Liquids and Gases

20.1 Liquids which conduct

As with solid materials, some liquids are capable of conducting electricity, while others are not. Generally speaking, the strong acids (sulphuric, hydrochloric and nitric) together with solutions of their salts in water (such as copper sulphate, silver nitrate, sodium chloride), and the alkalis (e.g. caustic soda, potassium hydroxide) are good conductors. On the other hand, the organic liquids (e.g. benzene, oils, alcohol, phenol) are poor conductors. A simple arrangement, as in Figure 20.1, can be used to test the conductivity of a liquid. Note the inclusion of a lamp in the circuit, important in case the electrodes touch one another and cause a short-circuit.

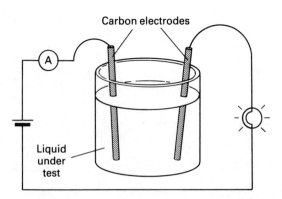

Fig. 20.1 Testing conduction through liquids

20.2 Electrolysis

The most important effect caused by a current in a conducting liquid is that chemical changes occur at the *electrodes*. The name given to this process is *electrolysis* and the liquid is called the *electrolyte*. The electrode where the (conventional) current enters the liquid is known as the *anode* and that at which it leaves the *cathode*. The anode would be connected to the positive terminal of the source of electricity and the cathode to the negative. A simple electrolytic cell, or voltameter, would have only one anode and one cathode, but an industrial one might have over twenty of each. Figure 20.2 shows two basic laboratory electrolysis experiments, and the table summarises the effects in a few cases. It will be seen that the material of the electrode is important, as well as the electrolyte.

Electrolyte	Electrodes Anode	Cathode	Action at Anode	Action at Cathode
Copper sulphate solution	Copper	Copper	Copper dissolves in electrolyte	Copper deposited from electrolyte
Copper sulphate solution	Platinum or carbon	Platinum or carbon	Oxygen liberated	Copper deposited from electrolyte
Dilute acids, e.g. hydrochloric	Platinum or carbon	Platinum or carbon	Oxygen liberated	Hydrogen liberated
Molten sodium chloride	Graphite (carbon)	Iron	Chlorine liberated	Sodium deposited
Molten aluminium ore	Carbon	Carbon	Oxygen liberated	Aluminium deposited

20.3 Electroplating and refining

The deposition of metals from solutions of their salts during electrolysis – e.g. copper from copper sulphate solution – has an

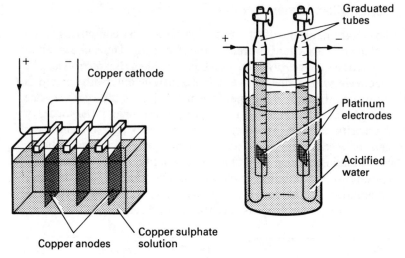

Fig. 20.2 Copper and water voltameters

immediate application in the electroplating industry. A thin metallic film can be formed on almost any shape of almost any non-porous material by making the object the cathode in an electrolytic cell. Copper, nickel and chromium plating, even of plastic objects such as car windscreen frames, have been increasingly used for decorative or protective purposes. The mark EPNS on cutlery indicates that the items are made of electroplated nickel silver, rather than solid silver or stainless steel.

Metals can be *refined* (made purer) by a similar process in which an impure anode and a pure cathode of the same metal are used. The cathode gradually increases in mass as pure metal is deposited on it from the anode via the electrolyte, gaining perhaps two hundred times its original mass in two weeks of electrolysis. Copper, zinc, tin, gold and silver can be refined in this way after their ores have been extracted in impure forms from the earth.

As well as electrolytic refining, metals can be extracted from their ores on a commercial basis by the electrolysis of fused (molten) salts excavated from the earth. Sodium and aluminium are both produced by electrolysis.

20.4   The mechanism of conduction in liquids

In metals the electric charges flowing through them which constitute a current are carried by electrons (Chapter 1), some of which are thought to be able to move almost as free particles within the metal. The electrons each carry a small negative charge of electricity. In conducting liquids the charges are carried by much larger particles, called *ions*, which are parts of atoms or molecules. An electrolyte such as fused sodium chloride is thought to be totally ionic, there being sodium ions (+ve) and chlorine ions (−ve). The potential difference between the electrons causes the ions to move, the positive ones being attracted to the cathode and the negative ones to the anode. Both the +ve and −ve ions are responsible for the conduction of the electric current (Figure 20.3).

The effects can be described as the sodium ion Na^+ and the chlorine ion Cl^- respectively gaining and losing an electron e^- at the cathode and anode:

$Na^+ + e^- \rightarrow Na$ (sodium atom released at cathode)
$Cl^- \qquad \rightarrow e^- + Cl$ (chlorine atom released at anode)

This process is equivalent to an election being transferred from cathode to anode through the electrolyte, and the production of sodium and chlorine.

Matters become more complex when solutions in water are electrolysed (e.g. hydrochloric acid), because the water also is

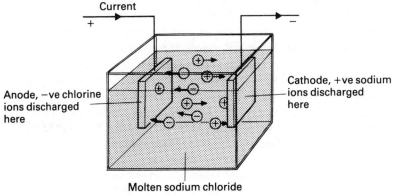

Fig. 20.3   Conduction by ions in molten sodium chloride

partially ionic. When there are two −ve ions present together, say, one will be preferentially discharged. (The order of precedence is known as the *electro-chemical series*, and the reader is referred to a book dealing with electro-chemistry for a full explanation of this effect.) Water ionises as H^+, OH^-, and H_3O^- and there would also be H^+ and Cl^- ions from the hydrogen chloride. The equations are not simple, but the net effect is the transfer of electrons from cathode to anode, together with the production of hydrogen and oxygen gases in the ratio 2:1 by volume.

20.5   Conduction through gases

Gases are not normally good conductors (if they were, all cells and batteries would quickly run down by continuous short-circuit through the air), but under very high voltages or at very low pressures they can be made to carry electric currents. Lightning is one extreme example of electricity flowing through a gas, whilst the familiar fluorescent tubes or neon signs are others.

Lightning is like an enormous spark between one cloud and another or between a cloud and the Earth which follows the build up of charge during a thunderstorm. Large amounts of energy are involved in a lightning flash and extremely high voltages (millions of volts) are developed.

Under normal circumstances there are a few air molecules in every thousand which exist as ions, but when a very high voltage is applied many more become ionised. The + and − ions are accelerated by the voltage, producing more ions by collisions with molecules until the air becomes a much better conductor than usual. The path taken by the current will zig-zag through the regions of highest ionisation, with heat and light being released in the process.

At low pressures ($\approx \frac{1}{100}$ normal atmospheric pressure) gases conduct more easily because the ions suffer fewer collisions and so can reach the necessary speed to produce further ionisation under much smaller voltages. The colour of the light emitted depends on the gas itself, an effect which led to the development of advertising signs consisting of narrow glass tubing with the appropriate low pressure gas for the required colour. Neon gives a red glow, helium a whitish blue, mercury vapour a bluish green, sodium vapour a bright yellow, etc.

Mixtures of gases can be found for almost any desired colour, and with narrow tubes the light is concentrated into a bright line of colour. The shape of the tube is immaterial. Apart from advertising signs, street lamps using mercury or sodium vapour are other examples of a gas discharge lamp, though these particular ones need a warming-up period first to produce sufficient vapour from the liquid mercury or solid sodium.

Fluorescent lights use the same technique except that the gas employed emits its energy mainly in the invisible ultra-violet region of the spectrum. The inside of the tube in this case is coated with a phosphor which absorbs the ultra-violet light and converts it to a mixture of visible colours. A close approximation to white light can be obtained in this way. The running costs of fluorescent tubes are much smaller for a given amount of light output than the ordinary filament lamp where significant amounts of heat energy are produced in addition to the light.

20.6 Sparks

Sparks across electrical contacts are another illustration of gaseous conduction. Often sparks are intentional, as in a piezo-electric gas lighter or a car sparking plug, for example, but sometimes they occur when contacts are made or broken, as in a light switch or in the vibrating contacts of an electric bell, which are not advantageous. The design of contact points needs to be considered carefully if useful energy is not to be wasted. In general, sharp corners or pointed shapes cause more sparking than flat or smooth surfaces (see section 14.9). Quite low voltages can be sufficient to produce sparks under the right conditions (such as the short-circuit of an accumulator), but as a rule they occur where potential differences are in the hundreds or thousands of volts or more. A capacitor (section 14.5) is a useful device for reducing unwanted sparking across moving contacts.

Since a spark or electrical discharge of any sort requires the presence of ions or other charged particles, a perfect vacuum would be a spark-free region.

20.7   Summary

Electrolysis occurs when currents flow through liquids.

Liquid conduction is through the movement of ions whose charges are neutralised at the electrodes.

A voltameter comprises the container, with electrodes and electrolyte, within which electrolysis takes place.

Electroplating, extraction and refining of metals use electrolytic processes.

Conduction through gases is possible at high voltages and low pressures, or when sparks occur.

Glossary of Technical Terms

a.c.   alternating current, the flow of charge backwards and forwards.
accumulator   a secondary source of electricity requiring charging before delivering current itself.
ammeter   an instrument for measuring current.
ampere   the unit of electric current.
ampere-turns   the product of current and number of turns on an electromagnet.
amplitude   the maximum value of an alternating quantity in one direction.
anode   the positive electrode in a voltameter.
anti-phase   a phase difference of 180°.
armature   part of a motor or generator on which the coils are wound, or the iron piece attracted by an electromagnet.

back e.m.f.   the reverse e.m.f. produced by electromagnetic induction, e.g. in motors.
base   the central element in a transistor.
battery   a group of cells connected in series.

capacitance   the property of a component defined by the ratio of charge to p.d., measured in farads.
capacitor   a component having capacitance.
cathode   the negative electrode in a voltameter.
cell   a primary source of electricity, usually chemical but can also be based on light or mechanical energy.
charge   the basic electrical property produced by rubbing materials; can be positive or negative, carried by electrons or ions; measured in coulombs.
circuit   a complete loop or network of connections linking components to a source of electricity.
collector   one element in a transistor.
colour code   used for electrical wiring or to denote resistor values.
commutator   a device for changing the direction of current in a d.c. motor, also found in a d.c. generator.
coulomb   the unit of electric charge.

current   a flow of charged particles such as electrons or ions, measured in amperes.

d.c.   direct current, flow of charge in one direction.
diode   a component that conducts only in one direction.
dynamo   a device for converting mechanical energy into electrical energy.

earth   used as a zero of potential and for safety connections for appliances.
eddy currents   currents induced in the cores of electromagnetic devices.
electrode   an electrical connection to a device.
electrolysis   the conduction of electricity through liquids.
electromagnet   a piece of iron with a magnetising coil around it.
electromagnetic induction   the effects produced by the interactions of moving magnetic fields and conductors.
electrometer   a device for measuring the effects of static electricity.
electron   a tiny particle present in all atoms; carries a negative charge.
electroscope   a device for showing the effects of static electricity.
e.m.f.   electromotive force, that which causes the movement of charges (electrons) around a circuit, e.g. by a cell; measured in volts.
emitter   one element in a transistor.

farad   the unit of capacitance.
frequency   the number of cycles (oscillations) or an alternating quantity per unit time, i.e. seconds, measured in hertz.
fuse   a thin wire which will melt at a certain current rating.

galvanometer   a sensitive instrument for indicating or measuring current or p.d.
generator   a device for converting mechanical energy into electrical energy.
grid   a distribution system for electricity.

henry   the unit of inductance.
hertz   the unit of frequency.

impedance   total effective 'a.c. resistance'; measured in ohms.
inductance   the property of a component to react against changes of current; measured in henries.
induction   the production of effects in one material by the influences of another, particularly magnetic or electrical.
inductor   a component having inductance.
insulator   a material which does not conduct electricity.
integrated circuit (IC)   a single piece of semiconducting material containing several ready-made circuits.
internal resistance   electrical resistance inside a cell, battery or power pack.
ion   a part of an atom or molecule with more or fewer electrons than normal; can carry positive or negative charge.

joule   the unit of energy or work.

kilowatt-hour the unit of energy equal to 3 600 000 joules.

lag and lead the amount by which one alternating quantity is behind or ahead of another.

Lenz's law the relationship between the direction of induced currents and the changes producing them.

logic gate a circuit whose output reflects the relationship between its inputs.

magnetic field the region of influence around a magnetised material or electric current.

magnetic materials those substances capable of being magnetised.

modulation the combination of a wave of one frequency with one of a much higher frequency.

multimeter a single instrument capable of measuring current, p.d. and resistance over a variety of ranges.

multiplier a resistor used to convert a galvanometer into a voltmeter.

n-type a semiconducting material in which the majority of charge carriers are negatively charged.

ohm the unit of resistance.

ohmmeter an instrument for measuring resistance.

Ohm's law the relationship between current and potential difference for a metallic conductor at constant temperature.

oscilloscope (CRO) an instrument used for measurement and display of fluctuating voltages.

parallel connection of components such that the same potential difference is applied to each of them.

period the time taken for one complete oscillation of an alternating quantity; measured in seconds.

phase angle phase difference; measured in degrees.

phase difference the amount by which two alternating quantities are out of step with one another.

potential electrical level, measured in volts.

potential difference (p.d.) difference in electrical level between two points; measured in volts.

power rate of working or rate of transfer of energy; measured in watts.

power pack a piece of equipment with definite outputs of current or voltage.

preferred values a set of values used for the manufacture of resistors.

p-type a semiconducting material in which the majority of charge carriers are positively charged.

reactance the frequency-dependent part of an a.c. impedance; measured in ohms.

receiver a system which responds to radio waves.

rectifier a device that converts a.c. to d.c.

resistance   the property of conductors which hinders the passage of current; measured in ohms.

resistor   a component having resistance.

resonance   the response of an oscillatory system to particular frequencies which cause very large amplitude.

ring main   loop of cable forming part of a house wiring system.

r.m.s. value   root mean square value, the d.c. equivalent of an a.c.

rotor   the rotating part of a generator.

semiconductor   a material with electrical properties intermediate between those of conductors and insulators, used as the basic material for diodes, transistors and other electronic components.

series   connection of components such that the same current passes through each of them.

short-circuit   a connection of negligible resistance across a cell or component.

shunt   a resistor used to convert a galvanometer into an ammeter.

solenoid   a cylindrical-shaped coil of wire, often used for magnetisation.

spur   an additional connection to ring main.

stator   the stationary part of a generator.

thermistor   a temperature-sensitive resistor, usually made of semiconducting material.

thyristor   a four-element semiconducting device.

time-constant   the time taken by an exponential change to reach 63 per cent of its final value.

transducer   a device which converts one type of signal into another, usually electrical.

transformer   a device which can change alternating voltages from one level to another, step-up or step-down.

transistor   a three-element semiconducting device, npn, or pnp junction type; or field effect type.

transmitter   a system which transmits radio waves.

Van de Graaff generator   a machine which generates a high static voltage.

volt   the unit of potential, potential difference or e.m.f.

voltameter   the vessel with electrodes and electrolyte in which electrolysis occurs.

voltmeter   an instrument for measuring p.d. or e.m.f.

watt   the unit of power.

Key Formulae and Relationships

An ampere is a rate of flow of a coulomb per second.
At any junction in a circuit, total current in equals total current out.
Ohm's Law states: current is directly proportional to p.d.

Resistance $R = \dfrac{\text{p.d.}}{\text{current}}$ or $R = \dfrac{V}{I}$

Resistors in series: $R = R_1 + R_2 + R_3$

Resistors in parallel: $\dfrac{1}{R} = \dfrac{1}{R_1} + \dfrac{1}{R_2} + \dfrac{1}{R_3}$

Power is rate of doing work or rate of transfer of energy.

Power, $P = IV = I^2R = \dfrac{V^2}{R}$

Capacitance $= \dfrac{\text{charge}}{\text{p.d.}}$ or $C = \dfrac{Q}{V}$ $Q = CV$

E.m.f. $E = IR + Ir$

Terminal p.d., $V = E - Ir$

Induced e.m.f. is proportional to the rate of change of magnetic flux.

Direction of induced current is such as to oppose the change producing it.

For a transformer $\dfrac{V_1}{V_2} = \dfrac{n_1}{n_2}$

Capacitors in series $\dfrac{1}{C} = \dfrac{1}{C_1} + \dfrac{1}{C_2} + \dfrac{1}{C_2}$

Capacitors in parallel $C = C_1 + C_2 + C_3$

Energy stored in a capacitor $W = \frac{1}{2}CV_2 = \frac{1}{2}QV = \frac{1}{2}\dfrac{Q^2}{C}$

Time-constant for capacitance-resistance circuit $= RC$

Time-constant for inductor-resistor circuit $= \dfrac{L}{R}$

Inductance, $L = \dfrac{\text{Back e.m.f.}}{\text{rate of change of current}}$

Energy stored in an inductor, $W = \frac{1}{2}LI^2$

Frequency and period are reciprocals of one another.

In a.c., current leads p.d. in capacitative circuits.

In a.c., p.d. leads current in inductive circuits.

$(\text{Impedance})^2 = (\text{Resistance})^2 + (\text{Reactance})^2$

Resonant frequency of capacitor-inductor circuit, $f_\mathrm{r} = \dfrac{1}{2\pi\sqrt{LC}}$

Root mean square value for a sine wave $= \dfrac{1}{\sqrt{2}} \times \text{amplitude}$

$$= 0.71 \times \text{amplitude}$$

Index